Essential Maths for Engineering and Construction

Don't let your mathematical skills fail you! We all make mistakes; who doesn't? But they can be avoided when you understand why you make them. In engineering, construction and science exams, marks are often lost through carelessness or from not fully understanding the maths involved. When there are only a few marks on offer for a part of a question there may be full marks for a right answer and none for a wrong one, regardless of the thought that went into answering.

Taking mistakes that undergraduate students often make as a starting point, this book not only looks at how you can prevent mistakes but provides an introduction to the core mathematical skills required for your degree discipline. The book contains many practice questions and fully worked answers, so whether you struggle with algebra, geometry, statistics, calculus or any of the other mathematical areas vital to your degree, this book will guide you around the pitfalls.

Focusing on the recognition, correction and avoidance of common errors to improve the clarity of your maths work and mathematical understanding, this book is for any undergraduate who does mathematics—and sometimes makes mistakes.

Mark Breach is Principal Lecturer in Engineering Surveying at Nottingham Trent University, UK. He teaches mathematics courses to civil engineering and architectural technician undergraduates.

Essential Maths for Engineering and Construction

How to avoid mistakes

Mark Breach

Spon Press
an imprint of Taylor & Francis

First published 2011
by Spon Press
2 Park Square, Milton Park, Abingdon, Oxon OX14 4RN

Simultaneously published in the USA and Canada
by Spon Press
711 Third Avenue, New York, NY 10017

*Spon Press is an imprint of the Taylor & Francis Group,
an informa business*

British Library Cataloguing in Publication Data
A catalogue record for this book is available from the British Library

Library of Congress Cataloging in Publication Data
Breach, M. (Mark)
Essential maths for engineering and construction : how to avoid mistakes / Mark Breach.
p. cm.
1. Mathematics–Textbooks. I. Title.
QA37.3.B74 2012
510.24′62–dc22 2011008536

ISBN 13: 978-0-415-57926-1 (hbk)
ISBN 13: 978-0-415-57927-8 (pbk)
ISBN 13: 978-0-203-85173-9 (ebk)

Typeset in Sabon
by Glyph International Ltd.

MIX
Paper from responsible sources
FSC www.fsc.org FSC® C004839

Printed and bound in Great Britain by
TJ International Ltd, Padstow, Cornwall

Contents

Part I

Finding the pitfalls

1 Introduction

New ideas build on previous ideas—you can't expect to get it all at once

Mathematics is a very different kind of subject from most of the others that you study. There aren't many essays to write and there isn't an awful lot to learn, but there is a great deal to understand. True, there may well be some formulae that you are expected to remember and you have to know what your calculator can do and how to use it. You may also be required to reproduce some geometric or trigonometric proofs. There certainly is no great feat of memory required to recall facts and figures, as you have had to do for many other subjects. What makes mathematics so different are the many different levels of understanding that you are required to have about the subject.

So, for example, to understand how to differentiate $x^3 \sin x$, a product of two functions, you need to understand the product rule of differentiation. To differentiate the individual terms in the product you will need to understand how to differentiate a polynomial and how to differentiate a trigonometric function. To understand the proof that $\frac{d}{dx}(\sin x) = \cos x$ means that you will have to understand the geometry of compound angles necessary to derive the trigonometric identity $\sin(x + \delta x) = \sin x \cos \delta x + \cos x \sin \delta x$, and how this leads to $\frac{d}{dx}(\sin x) = \cos x$. This in turn means that you need an understanding of trigonometric ratios; that is what $\sin x$ and $\cos x$ really are. To make sense of the concept of $\sin x$, where x is a variable, you need a good grounding in basic algebra and to achieve that you will have to be able to work confidently with arithmetic. The point of this rather long paragraph is that mathematics is concerned with logic and that one concept needs to be mastered before you can go and make use of it in developing a new concept; all rather daunting stuff.

Each new subject requires that you fully understand the previous subject because in mathematics one subject builds on others. For most people this can be a complicated process. For many it is also rewarding because of the feeling of satisfaction that goes with intellectual achievement.

It is doubtful whether there are any mathematics text books that you can just pick up and read cover to cover like a novel. Mathematics is often best tackled in small manageable chunks and most text books are laid out in that way, deliberately, so you can read, absorb, understand, practise to aid learning, and confirm your achievement.

For many, there are the occasional hurdles that seem insurmountable. Often a good teacher will be able to help you find other ways to understand the subject and so make progress. Occasionally, you may have to be content to learn *how* to do a particular piece of mathematics and the understanding of *why* it should be so comes later.

Why the bare minimum isn't

If your mathematics is only formally assessed by examination you may be tempted to leave it until just before the exam with the intention of mugging it all up in the week or so just before. Many students only work to the next deadline and, with course work in other subjects seeming to be more pressing, it is often tempting to leave the more challenging subjects until later. However, since mathematics is a conceptually hierarchical subject it takes time to make sense of it.

Mathematics is usually best approached on a little-but-often basis. One hour a day is much more likely to be productive than a whole day once a week. But if it is to be little, then it must be often.

It looks so easy when the lecturer explains it

There are some lecturers who have the rare skill of making the complicated seem easy to understand. In mathematics that often means explaining one step so that it is well understood and then going onto the next step, explaining that well, and so on. However, in mathematics, understanding the whole is more than understanding each of the parts. To understand the whole you need to be able to put all the parts together. Following a proof is one thing; understanding it in both your head and your heart is quite another.

And this is why going to the lectures is not enough. The lectures merely give an introduction to the subject; they acquaint you with its terms and their relationships, but of themselves are unlikely to give you much understanding. Understanding mathematics is more likely to come with practice. If you get it right first time that is great, but often you will not. It is said that the best learned lessons are the most expensive; expensive in time, that is. Making mistakes and learning from them can be a very effective aid to understanding, as long as you don't make the same mistakes again of course.

Here is the motivation for this book. We all make mistakes; who doesn't? And when we discover what they are, we often call them 'silly'. But many of the mistakes can be avoided by knowing the reasons for the mistakes in the first place and the way they are often made.

Perhaps you find mathematics a difficult subject and, although you can do the maths by following instruction, it may be that you do not always understand what it is that you are doing. This book examines many of the common errors that students make and shows how they may be avoided.

How to avoid mistakes

This introductory chapter discusses the nature of mathematics from your point of view and gives advice on how to tackle the subject. The next chapter is more specific and discusses ways in which checks can be made to spot and hence correct mistakes.

Part II, *Misleading with mathematics*, is about the use and abuse of mathematics, mostly the latter. It discusses how mathematics is often misused, either deliberately or through ignorance. Part III, *Some mistakes that we make*, presents many of the specific errors that students make, either because of ignorance or because of carelessness.

It's all Greek to me

Yes, indeed so. Some of it *is* presented in Greek; well just Greek letters actually, but when wrapped up in maths notation, at a first glance, it can all look quite formidable; both the Greek letters as well as the maths notation. So to help make sense of it all, here is a 'decode' list. It is likely that you have already come across π, pi, used almost exclusively as the ratio of the circumference to the diameter of a circle, and possibly the first few letters of the Greek alphabet, α, β and γ.

The Greek alphabet

α A	alpha	η H	eta	ν N	nu	τ T	tau
β B	beta	θ Θ	theta	ξ Ξ	xi	υ Υ	upsilon
γ Γ	gamma	ι I	iota	o O	omicron	ϕ Φ	phi
δ Δ	delta	κ K	kappa	π Π	pi	χ X	chi
ε E	epsilon	λ Λ	lambda	ρ P	rho	ψ Ψ	psi
ζ Z	zeta	μ M	mu	σ Σ	sigma	ω Ω	omega

Basic mathematical symbols

$+$	add	$>$	greater than
$-$	subtract or negative sign	$<$	less than
$\times$	multiply	$\geq$	greater than or equal to
/ or $\div$	divide	$\leq$	less than or equal to
$=$	equals	$\pm$	plus or minus
$\approx$	approximately equal	$\mp$	minus or plus
$\equiv$	identical	∞	infinity
$\neq$	does not equal	∂	partial differential

δ	infinitesimal change	%	percent
Δ	finite change	$\surd$	square root
$\prod$	product	$\int$	integral
$\sum$	summation	$\therefore$	therefore
!	factorial	$\because$	because
^	power	$\propto$	proportional to

Your first maths result; relief or despair

Universities and colleges often start their mathematics teaching at what can appear to be a deceptively easy level. The reasons for this may include a desire not to alienate the less confident at the start of their course. If there is a mixed ability group the tutors may be seeking, initially, to bring all the class up to the same level before progressing onto the mathematics necessary to support other teaching. If this is the case and you feel the work is easy because you have done it all before, do not be lulled into a false sense of security and then later find that you have been left behind because you were not paying sufficient attention.

Use any initial tests, exams or other individual assessments to gauge your own knowledge in comparison to that of your peers. If you are well above the average of your group take comfort, for now. If you are well below, then use the experience as a wake-up call that you need to increase your efforts.

In examinations, marks are more often lost through carelessness than through ignorance. When there are only a few marks on offer for a part of a question, there may be full marks for a right answer and none for a wrong one, no matter that there was some merit in the thought that went into working out the answer.

Mathematics examination questions tend to be quite short in comparison to questions in some other subjects. Usually, the questions are rich in information and instruction and it is essential to read the question more than once to be sure of getting all the critical information from it. For example, if the question was:

> Given that 1 inch = 25.4 mm, what is 3 ft^2 in square metres to 2 decimal places?

This simple question contains some, but not all, the necessary information to enable you to perform the calculation. Although you have a link between imperial and metric linear measure you still need to know how many millimetres there are in a metre, 1000, and how many inches there are in a foot, 12. You also need to realise that the question is concerned with area, not linear measure. Finally, there is instruction on the level of precision that is required for your answer; you need to be sure you know the difference between decimal places and significant figures.

There is only one right answer to this question, as is usually the case in maths. It is 0.28 m^2. Anything else such as 278709.12 mm^2, 0.27 m^2, 0.278 m^2, 0.28 m or 0.28 does not meet the requirements of the question, either because it is wrong or because there is an absence of units, the wrong level of precision or incorrect rounding.

Often it is useful, sometimes essential, to draw a diagram to be able to understand the question. For example, if the question was:

> Solve the triangle ABC, where A is 23° 32′ 25″, a is 22.333 m and b is 34.444 m. Give angles to the nearest second and distances to the nearest millimetre.

This is obviously a problem of trigonometry and it is essential to be able to see the relationship between the known and unknown angles and sides. That will then indicate to you which trigonometric relationship to use; sine or cosine formula. But there is a further catch here which may not be apparent from the problem as expressed in words; there are two possible solutions to this problem.

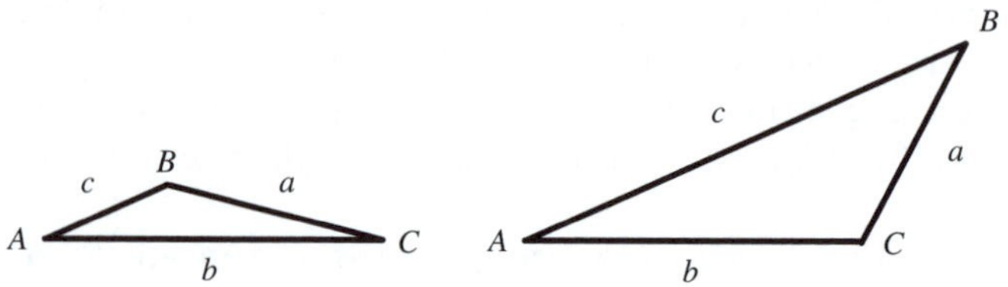

The two solutions are:

B 38° 1′ 24″ C 118° 26′ 11″ c 49.171 m and
B 141° 58′ 36″ C 14° 28′ 59″ c 13.985 m

Here is another problem, this time in mechanics:

> A man is working at the top of a 10 m vertical ladder when the ladder becomes detached. The foot of the ladder remains stationary. When the ladder has swung through an arc of 30° the man lets go and slides down a rough slope inclined at 30° to the horizontal. He arrives at the bottom of the slope, which is at the same level as the bottom of the ladder, just as he comes to rest. What is the coefficient of friction of the slope?

Although the description in the question is complete, it is very difficult to see what is going on here without a diagram. It is usually helpful to draw a diagram to make sense of the question; such as this one:

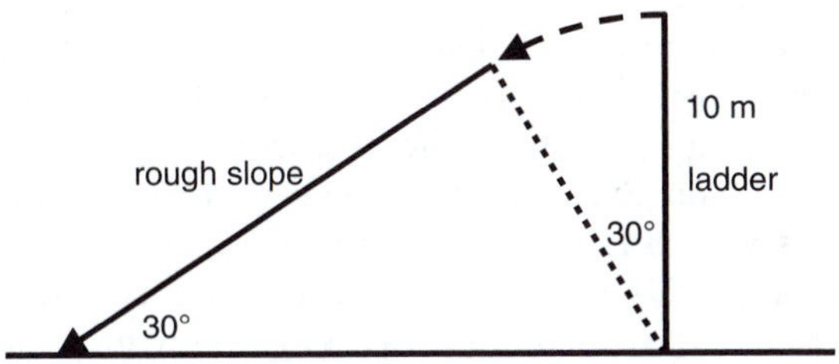

2 Find an independent check

Once again, we all make mistakes, if only because we are human and imperfect. However, mistakes can be costly, annoying and embarrassing in real life and in examinations they will cost you marks. There are a number of checks that can usually be applied to test the correctness of your answers. Some of the checks are approximate and although they will not confirm that your answer is correct they may show that the answer is clearly wrong; in which case that gives you the opportunity to investigate why, and hopefully re-examine your answer with a view to getting it right.

The worst kind of check is just to follow through your original working looking for errors. The problem with this approach is that you will tend to see what you expect to see, and as you are hoping that your answer is correct you are not likely to see the errors. A much better approach is to find a check that does not need reference to your original working. This way, your check is independent and much less likely to be corrupted by your expectations. Here are some suggestions.

Is the order of magnitude sensible?

Look at the size of your answer and ask yourself if it looks reasonable. If you have to work out 13% of £2345.67, would an answer of £30.49 ring any alarm bells? Just a quick glance at the numbers would lead you to expect the answer to be in the order of a few hundreds, not a few tens of pounds.

If the scenario set for the problem is one in which you can imagine the outcome that you are required to calculate, see if it fits within reasonable expectations. If the problem involves calculating the speed of a vehicle, would a speed of 867 kilometres per hour be likely? Would the speed of an aircraft flying at 12.5 kilometres per hour seem appropriate?

If the problem you have to solve relates to a real-world situation, consider whether the outcome is likely. To calculate the time of a transatlantic flight you might be given the distance between London and New York and the speed of the aircraft. If your answer comes out to be in minutes or days rather than hours you can be pretty sure you have made a mistake. If you need to calculate the fuel consumption of a car and you have measured the distance covered on a tank full of fuel, then you would expect the answer to be in the order of a few tens of miles per gallon or a few litres per hundred kilometres; anything vastly different would suggest an error in your calculations.

It is useful to know some very approximate physical relationships. In terms of imperial to metric conversions:

1 metre ≈ 10% more than 1 yard	1 yard ≈ 10% less than 1 metre
1 kilometre ≈ 0.6 miles	1 mile ≈ 1.6 kilometres
	1 furlong ≈ 200 metres
1 nautical mile ≈ 2000 yards	1 mile ≈ $\frac{7}{8}$ nautical mile
1 square metre ≈ 10 square feet	1 square foot ≈ 0.1 square metre
1 hectare ≈ 2.5 acres	1 acre ≈ 0.4 hectares
1 square kilometre ≈ 0.4 square miles	1 square mile ≈ $2\frac{1}{2}$ square kilometres
1 cubic metre ≈ 1.3 cubic yards	1 cubic yard ≈ $\frac{3}{4}$ cubic metre
1 litre ≈ $1\frac{3}{4}$ pint (UK)	1 pint (UK) ≈ 0.6 litre
1 litre ≈ 2 pint (US)	1 pint (US) ≈ $\frac{1}{2}$ litre
1 litre ≈ 0.22 gal (UK)	1 gallon (UK) ≈ $4\frac{1}{2}$ litre
1 litre ≈ 0.25 gal (US)	1 gallon (US) ≈ 4 litre
1 tonne ≈ $1\frac{1}{2}$% less than 1 ton	1 ton ≈ $1\frac{1}{2}$% more than 1 tonne
1 metre per second ≈ 2 miles per hour	1 mile per hour ≈ $\frac{1}{2}$ metre per second
1 kilogram ≈ 2.2 pounds	10 pounds ≈ $4\frac{1}{2}$ kilograms
	1 hundredweight ≈ 50 kilograms
1 kilowatt ≈ $1\frac{1}{3}$ horsepower	1 horsepower ≈ $\frac{3}{4}$ kilowatt

$1\ \text{lb/ft}^2 \approx 5\ \text{kg/m}^2$
$1\ \text{lb/ft}^3 \approx 16\ \text{kg/m}^3$

$1\ \text{kg/m}^2 \approx 0.2\ \text{lb/ft}^2$
$1\ \text{tonne/m}^3 \approx 62\ \text{lb/ft}^3$

Simple approximate sums give approximate answers

If you are presented with a problem with complicated numbers you can perform a quick check if you approximate all the input data and use that to get an approximate solution. For example, if the problem was:

> Calculate the cost of 478 bags of cement at £3.16 per bag.

You could approximate these quantities to 500 bags and £3 per bag. Your approximate cost would then be $500 \times £3 = £1500$. This is a sum you could (hopefully) do in your head. See how close it is to the correct answer of £1510.48.

In the problem:

> To four significant figures find the average speed of a truck that travels 368 kilometres in 5 hours 12 minutes.

Approximate these quantities to 350 kilometres and 5 hours. Dividing the distance by the time gives the approximate solution of 70 km/hr, which helps to confirm the true answer of 70.77 km/hr.

Put the answer back into the question

When you are given equations to solve, one useful check is to put all your answers back into the original equations to see if the original relationships are true. For example, with the simple linear equation:

$$5x + 7 = 22$$

you find the answer to be $x = 3$. Now replace x with 3 and see if the left-hand side of the equation comes out to be the same as the right-hand side.

$$5 \times 3 + 7 = 22$$

It does. This confirms that you have the right answer.

When dealing with simultaneous equations it is necessary to put your answers back into *all* the original equations. This is the only way to check that your full set of answers is correct. For example, you have these equations to solve for x, y and z:

$$2x + 3y + 4z = 20$$
$$4x + y + 3z = 15$$
$$3x + 2y + 2z = 13$$

and your solution is $x = 2$, $y = 4$ and $z = 1$.

It is insufficient to put the values back into just one or two of the equations. They must go into *all* the original equations, as the following illustrates:

$$2 \times 2 + 3 \times 4 + 4 \times 1 = 20 \quad \text{correct}$$
$$4 \times 2 + 1 \times 4 + 3 \times 1 = 15 \quad \text{correct}$$
$$3 \times 2 + 2 \times 4 + 2 \times 1 \neq 13 \quad \text{wrong}$$

It is quite possible to come up with a wrong answer that satisfies some but not all of the equations. The correct answer of course is $x = 1$, $y = 2$ and $z = 3$; you can check it for yourself.

When you are required to solve a quadratic equation there are either no real answers, one real answer or, most often, two real answers. So if you have found two answers it is necessary to confirm that they *both* satisfy the original equation. So, for example, if the equation was:

$$x^2 - 10x + 23.31 = 0$$

and your solution was that $x = 3.7$ and $x = 6.3$, then put them *both* back into the original equation:

$$3.7^2 - 10 \times 3.7 + 23.31 = 0 \quad \text{correct}$$
$$6.3^2 - 10 \times 6.3 + 23.31 = 0 \quad \text{correct}$$

Trigonometric checks

If you have computed the angles of a triangle, check that they add up to 180°.

If you have computed all the angles and all the sides of a triangle, check that the smallest side is opposite the smallest angle, the middle side is opposite the middle angle and the largest side is opposite the largest angle.

Try to build an independent check into your calculations. For example, if you are given the three sides of a triangle and need to calculate the three angles, then calculate each one independently using the cosine formula. Finally, check that the three angles do indeed add up to 180°.

Does the answer fit the diagram?

When trying to solve problems in trigonometry and geometry, always draw a diagram as accurately as possible with a protractor and a ruler. When you have completed your calculations of the distances and the angles, measure the same quantities on the diagram. If there is a large difference between what you calculate and what you measure, then either you have an error that needs to be investigated or you have missed a possible solution.

For example, if you are using the sine rule to find an angle, remember that given a positive value for $\sin x$ there are two solutions for x in the range 0° to 180°; one is 180° minus the other. This is the basis for the two possible triangles in the problem in Chapter 1.

Do the units of the answer match those expected from the question?

If a dimensioned quantity is to be calculated you can check that the units of the answer match the requirement of the question. For example, if you are given the surface area of a sphere in square metres and asked for the volume of the sphere, then you would expect the units of the answer to be in cubic metres. If you have the mass of an object in kilograms and its volume in cubic metres and are required to calculate its density, then you would expect the answer to be in kilograms per cubic metre. These are fairly trivial examples, but the problem becomes more complicated when you need to change units within the calculation. For example, you are given the change of speed of a vehicle in miles per hour over a period of time in minutes and are required to find the vehicle's acceleration in metric units. You need to make sure that your calculation leads you to a solution in metres per second squared.

Reverse the process

Calculus is not everybody's favourite subject and solutions are often hard to reconcile using any of the ideas above. If you have a problem that requires differentiation, you could always integrate your answer to see if it brings you back to the expression that was originally differentiated. Similarly, if integration is what you have to do, check your answer by differentiating it to confirm that you get back to the expression you originally integrated. For example, you perform the following integration:

$$\int e^x (\sin x + \cos x)\, dx = e^x \sin x + C$$

but do not feel confident about your answer. Use the product rule to differentiate your answer and that brings you back to the expression that you originally integrated.

$$\frac{d(e^x \sin x + C)}{dx} = e^x \sin x + e^x \cos x = e^x (\sin x + \cos x).$$

So the answer was right.

Part II

Misleading with mathematics

3 When ink meets paper

Does scruffy working indicate scruffy thinking?

Scruffy work certainly indicates no desire to work neatly. But is neatness necessary? If your work is neat, it is less likely that you will make an error due to misunderstanding your own writing. In maths you tend to write very little, much less than you would need in an essay for example, so the extra time taken to make your work neat is really very small. By making it neat you force yourself to think about what each number or symbol means and so may avoid sloppy thinking as well.

Lining up

Here is an addition sum. Can you see what went wrong?

```
12345
 9876
-----
13331
```

The correct answer is 22 221. The numbers to be added have not been lined up to make the units above the units, the tens above the tens, the hundreds above the hundreds, etc. In this case the units of 5 and 6 have been added correctly, but in adding the tens the position of the digit 8 in the lower number has been confused and it has been taken as a ten rather than a hundred. Similarly, with the 9 as a hundred, rather than a thousand. And so the 7 has been squeezed out and lost altogether.

Be careful of the vertical position of your numbers and symbols.

Consistent notation

What has gone wrong here?

Q. Expand the brackets $(2x + y^2)(x^2 + 3y)$

A. $(2x + y2)(x2 + 3y) = 2xx2 + 6xy + y2x2 + y23y$

$$= 4xx + 6xy + 4yx + 6yy$$

$$= 4x^2 + 10xy + 6y^2$$

The correct answer is $2x^3 + x^2y^2 + 6xy + 3y^3$.

By carelessly writing superscript indices as if they were ordinary numbers we have lost sight of which are the superscripts, and incorrectly multiplied the numerical coefficients by them.

Correcting mistakes

We all make mistakes, then spot them and make corrections. But crossings out and subsequent corrections may cause further error. Can you see what could go wrong here?

$436 \times 31 = 13316$

The answer as written was 13316 and subsequently corrected to 13516 but as written could be interpreted as either. Corrections should not be made by overwriting. Cross out the error and then write the correction neatly alongside.

$436 \times 31 =$ ~~13316~~ 13516

Diagrams that mislead

Graphs are an excellent way of making huge lumps of numerical data easy to understand. For example, a trader wishes to review the price of a product called Heavy Metal over a period of time. Here is the graph of recent prices. Notice how the *y*-axis has been truncated to emphasise the fluctuation of the price over the period depicted.

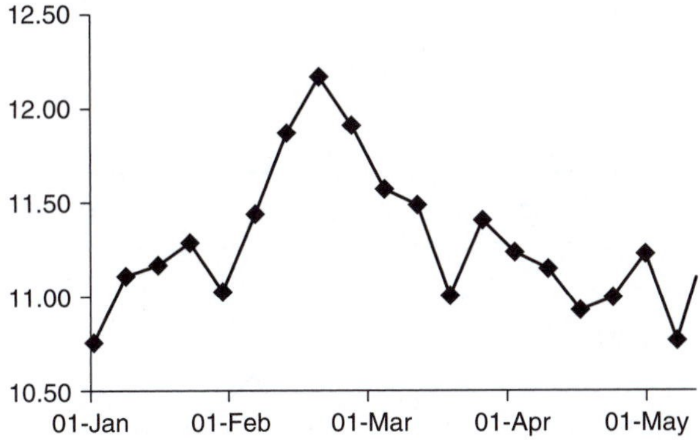

A better view of the trend, or in this case stability, would be if the full range of the *y*-axis were shown. There appears to be barely any significant movement at all here. However, only the average weekly prices are shown. Had the daily or even the hourly prices been shown quite a different story of the volatility of the price, as below, might have emerged.

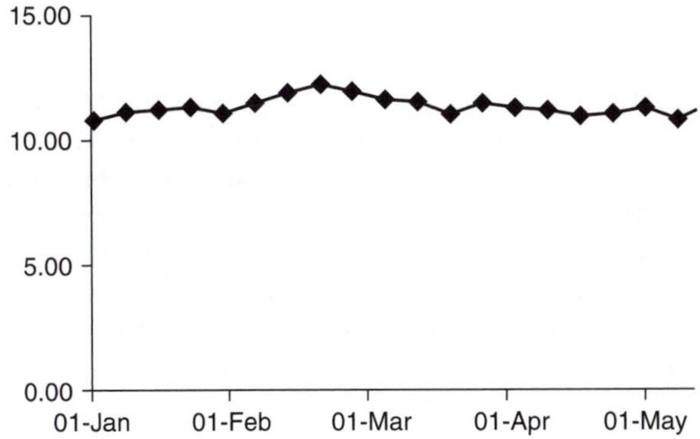

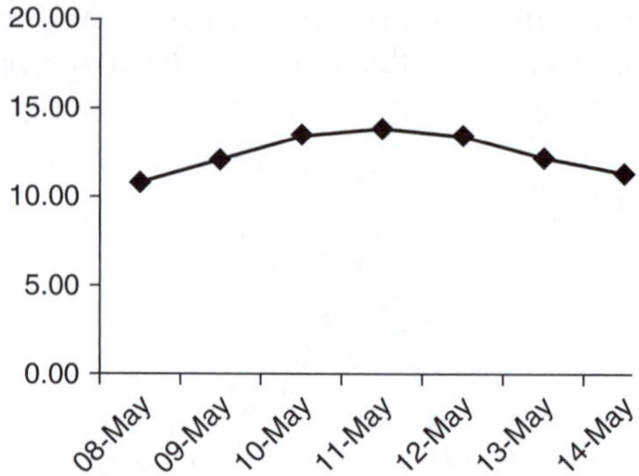

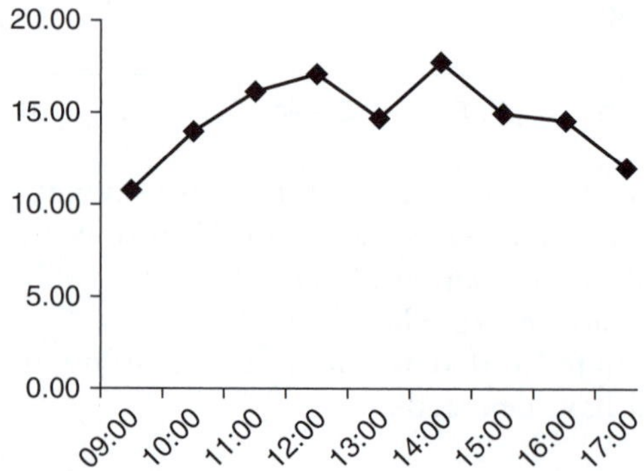

Graphs can be manipulated in other ways to persuade the reader to draw conclusions that are not supported by the data.

Ninety-nine people are asked to state their preference for different types of confectionery. The results are that 32 favour Gloo, 33 like Gulp and 34 go for Gush. Statistically there is no significant difference, that is, you cannot say with meaningful confidence that any one confection is preferred over any other. But here is how the advertising executive may have presented the data.

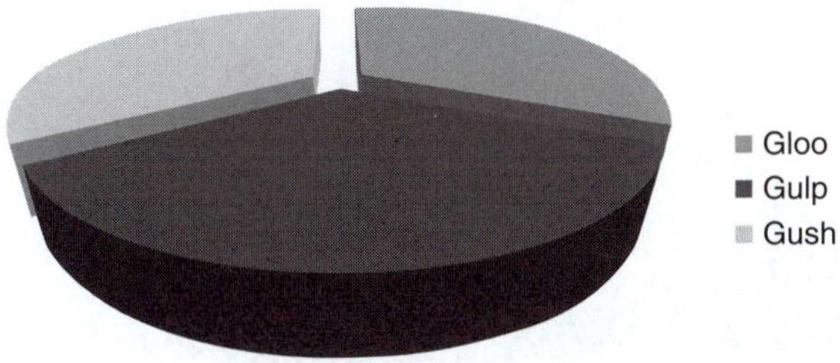

Here shading, perspective and rotation have been used to bias the conclusion. A wide-angle perspective has been used. Gulp is in a dark colour, its segment has been placed nearest the reader and that segment partly obscures those of the other products.

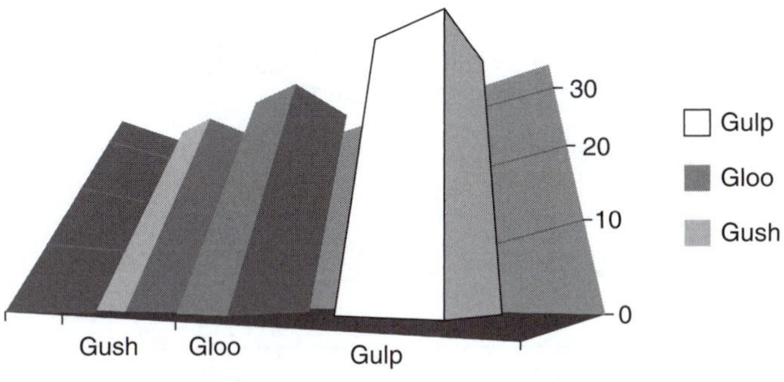

In this graph notice that the shameless use of rotation and perspective puts Gulp nearest the reader and makes it look much bigger than its rivals. In fact, against the scale, it looks much bigger than its value of 33 would suggest. See also how shading has been used to make the competitors merge with the background and how Gulp's black outline gives it greater clarity through contrast. In fact, by truncating the *y*-axis the effect can be made even more dramatic and also misleading.

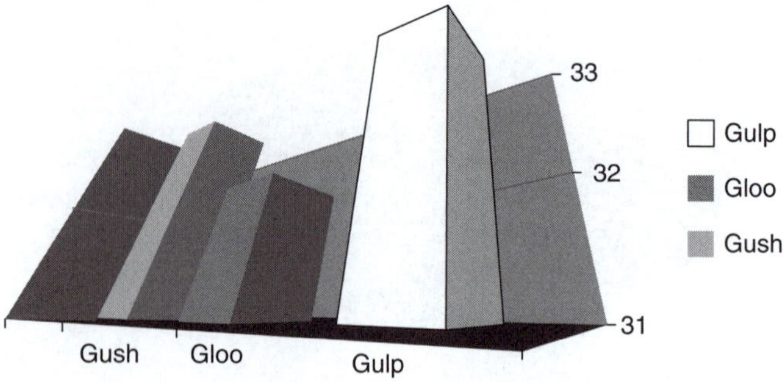

Something to think about

Which of the following graphs are misleading, and why?

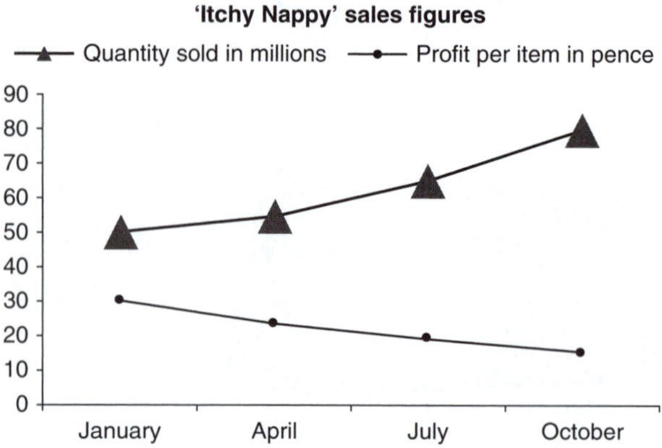

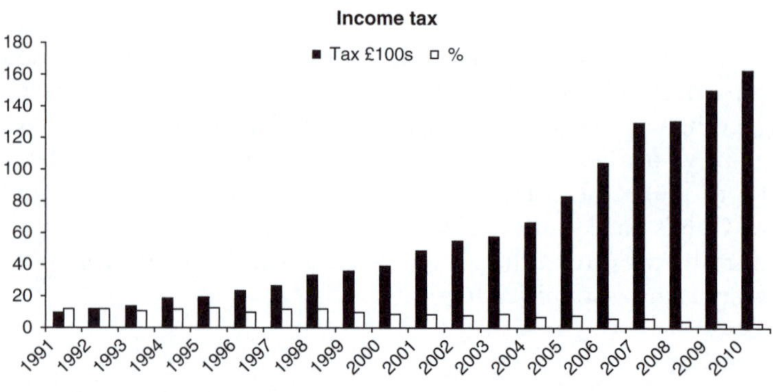

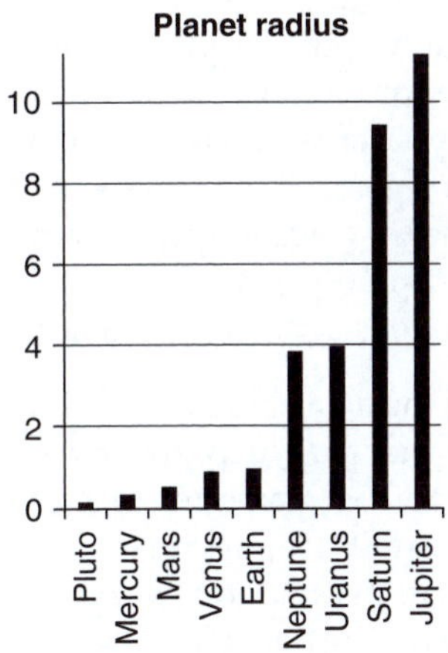
Planet radius
0
2
4
6
8
10
Pluto
Mercury
Mars
Venus
Earth
Neptune
Uranus
Saturn
Jupiter

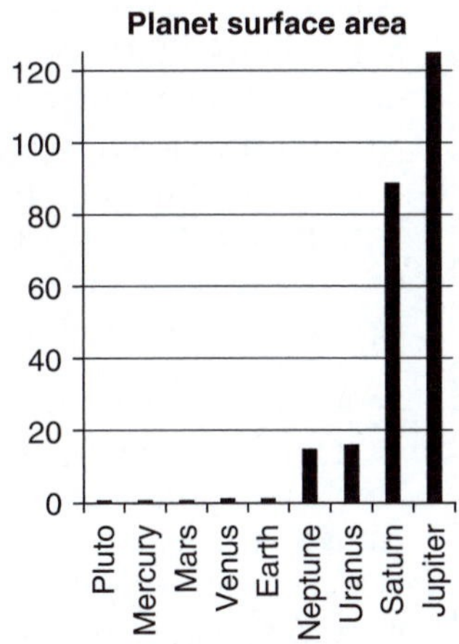
Planet surface area
0
20
40
60
80
100
120
Pluto
Mercury
Mars
Venus
Earth
Neptune
Uranus
Saturn
Jupiter

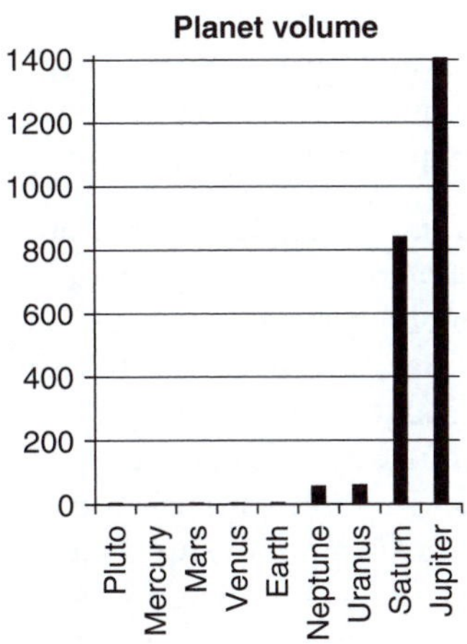
Planet volume
0
200
400
600
800
1000
1200
1400
Pluto
Mercury
Mars
Venus
Earth
Neptune
Uranus
Saturn
Jupiter

Itchy Nappy. There are two lines on the graph. The positive slope for the number of sales is steeper than the slope representing the decline in the profit. However, this graph looks rather as if the sales manager is trying to bury bad news under seemingly good news. True, the sales figures have gone up from 50 million to 80 million but the profit per item comes down from 30p to 15p per item. So if we put the two figures together then in April the profit was £0.30p × 50 000 000 = £15 000 000 but in October it was £0.15p × 80 000 000 = £12 000 000; a decline of 20%. So notice the emphasis on the positive leading to suppression of the negative.

Income tax. We all grumble about the tax we have to pay but it is very difficult to avoid without attempting evasion. If you try, the taxman will have something to say about that. Here is my friend's tax bill for the last 20 years. He complains more bitterly than most; but should he? Certainly, his tax bill (the black bars) rises year on year but hidden at the bottom in the white bars is the percentage of his income that he pays in tax which, because of his astute accountant, tends to get less every year. In fact, putting the statistics together, they show that his untaxed income is rising at an extraordinary rate: from £7000 in 1991 to £524 000 in 2010. Notice again how the choice of scale buries the good news. He should grumble!

Planets. The y-axes of the three graphs representing the planets are scaled by the earth, and so are in terms of earth radii, earth surface areas and earth volumes, respectively. So which one correctly represents the relative sizes of the planets? The answer is that they all do. It depends on how you define size; by radius, surface area or volume. That in turn would depend upon your intended application of the data, that is, the message that you are tying to get across with your graph.

Predictive models that miss the obvious or predict the impossible

It is very tempting to draw conclusions from apparently reliable data. Imagine that you decided to stand for election as an independent member of parliament, an MP. Being unknown, in your first election very few votes were cast for you; you got 10% of the vote. Over the years, as you became known locally, your share of the vote increased; at each election your share of the vote increased by exactly 10%. At the last election you got 40%. So what will be your outcome in 2014? And again in 2038, assuming an election every 4 years?

Year	*Election relative to 2010*	*% of vote*
1997	–3	10
2001	–2	20
2005	–1	30
2010	0	40
2014	1	?
2018	2	?
2022	3	?
2026	4	?
2030	5	?
2034	6	?
2038	7	?

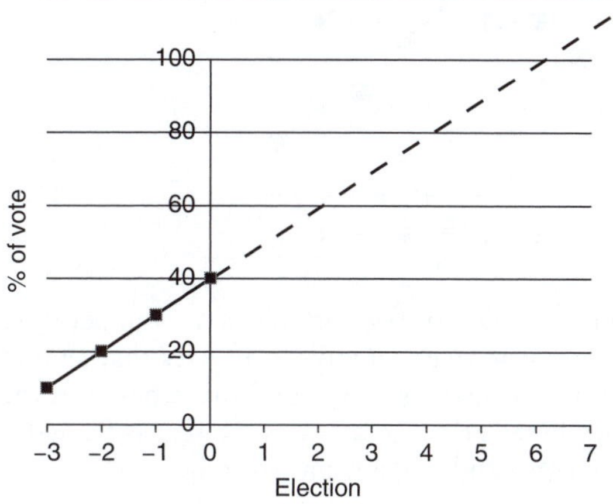

The steady rise in your popularity is a historical fact but can a 'law' be established from it? If it could it would guarantee a win at the 2014 election; surely impossible given the nature of politics. Even more dramatic is your share of the vote in 2038; 110%!

The error here was in assuming that a straight-line graph, which is valid over a given range, is valid over any extension to that range.

Hooke's law states that when a force is applied to stretch an object such as a spring, the extension of the object is proportional to the force. So if you double the force you would expect to double the extension. The law is correct but only within its valid range. Once the object reaches its 'elastic limit' it moves into a plastic state and subsequently breaks.

Here are the monthly average temperature records for two towns just 10 miles apart.

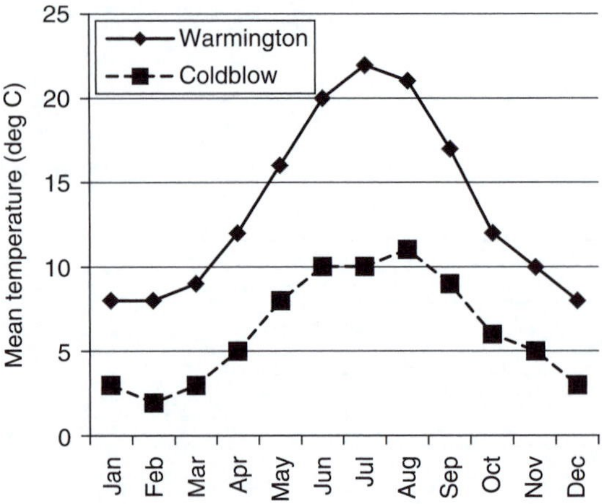

Although the graph shows a big difference in temperature between these two towns, it gives no indication why that should be so. In fact, one town is near the coast and the other is in the mountains. The point here is that data out of context is flawed if you intend to draw meaningful conclusions from it.

In an experiment to predict tidal levels at an anchorage, you observe the depth of water in metres over a period of 3 hours, at half hourly intervals. Here are your results.

Time	*Depth*
07:00	6.63
07:30	7.24
08:00	7.64
08:30	7.80
09:00	7.70
09:30	7.36
10:00	6.80

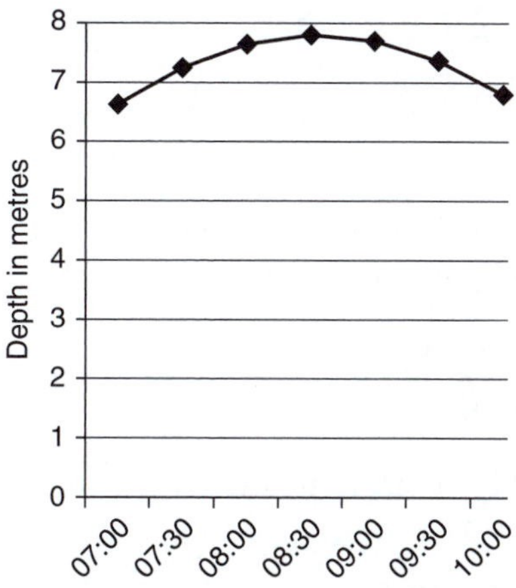

You note that the average depth is well over 7 metres and since the draught of your boat is only 1 metre you confidently conclude that it is safe to anchor here. You do so and proceed for a night ashore. However, if you had plotted the data on a graph you would have noticed that all your depth measurements were taken within 90 minutes of the top of the tide.

Since the tide approximately follows a sine curve, you could expect the tide to drop considerably further. Had you modelled the data over a complete 12½ hours, the approximate tidal period, as below, you would have concluded that you were going to run aground, which given that your boat has a single fin keel would mean that it would fall over and flood on the next rising tide. The point is that you can miss the obvious by making predictions based upon incomplete data or incorrect modelling resulting from the limited data that you have gathered.

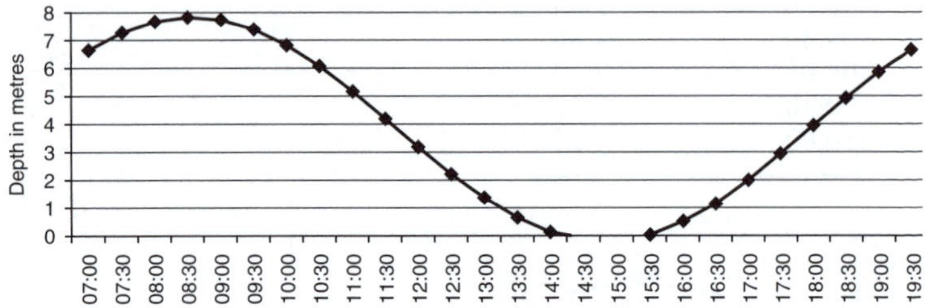

Continental notation

You are probably quite happy that you recognise numbers when written. For example, you would probably agree that

is ten point zero zero seven. However, to many Europeans the number 10.007 would be written as:

Notice the different ways that the figures 1 (one) and 7 (seven) may appear in handwriting; in particular a European 'one' is remarkably similar to a British 'seven'. Notice also the different usage of the dot and comma. In British script a dot indicates a decimal separator (decimal point) and a comma is used as a delimiter or spacer. In European usage it is the other way around;

a comma is used as a decimal separator and a dot may be used as a delimiter.

Something to think about

Can you unambiguously identify the following numbers?

1. 234.565,432,345
2. 234,565.432.345
3. 234.565,432
4. 234,565.432
5. How can the dot and comma ambiguity be resolved?

---

1. Yes. There are two commas and one dot so the dot must be the decimal separator, because there can only be one, and the commas must be delimiters.
2. Yes. There are two dots and one comma so the comma must be the decimal separator and the dots must be delimiters.
3. No. This could be two hundred and thirty-four point five six five four three two in British notation ***or*** two hundred and thirty-four thousand five hundred and sixty-five point four three two in European notation.
4. No. This could be two hundred and thirty-four thousand five hundred and sixty-five point four three two in British notation ***or*** two hundred and thirty-four point five six five four three two in European notation.
5. In modern notation the delimiter or spacer is represented by a thin space. This leaves only one symbol in the digits of a number and, whether it is a dot or a comma, it represents only the decimal separator.

So 123 456.789 is the same as 123 456,789.

Selecting subsets to bias results

When you conduct an experiment you do not always get the results you expect. For example, perhaps you have been investigating Hooke's Law which is concerned with how the extension of a sample of material varies with the force applied to it. Hooke's Law states that the extension is proportional to the force applied, providing the material does not exceed its elastic limit.

$F = kx$ where F is the force applied
k is a constant for the sample
x is the extension of the sample

Suppose you have successfully shown that this is true for samples of steel, copper, wood and glass and found the constant for each sample, k, which is the slope of the graph, but now wish to investigate rubber. You set up the experiment and take the following readings and so construct this graph.

Force newtons	*Extension metres*
0	0.000
10	0.004
20	0.010
30	0.020
40	0.034
50	0.050
60	0.066
70	0.080
80	0.090
90	0.096
100	0.100

This curve does not fit with the straight-line shape of the graph you were expecting, so how can you find k, the constant for this sample? You choose any two data points to find the slope by dividing the difference in force by the respective difference in extension.

Using the first two points you get $k = \frac{10-0}{0.004-0.000} = 2500$ newtons per metre and using the last two points you get

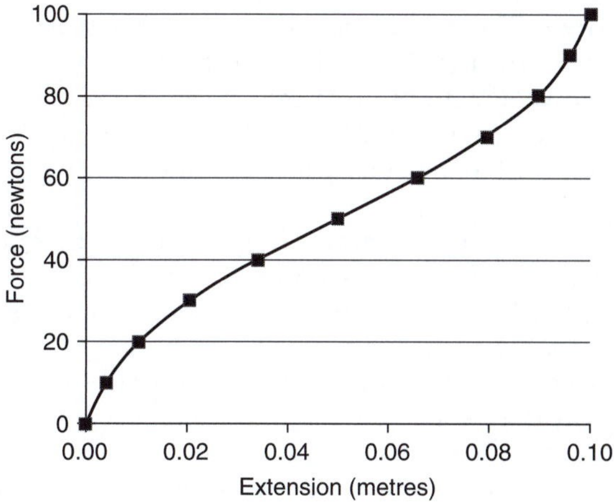

$k = \frac{100-90}{0.100-0.096} = 2500$ newtons per metre. And these two results agree. However, using two points from the middle $k = \frac{60-50}{0.066-0.050} = 625$ newtons per metre.

This is substantially in disagreement; so what should you do? It is tempting to ignore some of the data. If you do, what validity will your results have? The conclusion you can get from this is that the sample appears not to follow Hooke's Law. This is a far more appropriate conclusion than any that can be drawn by being selective with your experimental data. This is a conclusion that you may not have been expecting so you should test your new hypothesis that 'rubber does not follow Hooke's Law' with other experiments to confirm that this is so and that your result is not just an error in your measurements or computations, or a fault with your experimental apparatus.

Disregard for dimensions

Here is a 'proof' that 1 metre is the same as 1 centimetre. Can you spot the mistake?

$$1 \text{ m} = 100 \text{ cm} = (10 \text{ cm})^2 = (0.1 \text{ m})^2 = 0.01 \text{ m} = 1 \text{ cm}$$

How long did it take you to spot the mistakes; actually there are two. The first comes from assuming that 100 cm = $(10 \text{ cm})^2$. It is true that $100 = 10^2$ but the dimensions have been corrupted. A centimetre is not a square centimetre; one is linear, the other is area. Likewise, $(0.1 \text{ m})^2 = 0.01$ m is not correct. Although $0.1^2 = 0.01$, a square metre is not the same as a linear metre.

A number is just a number. If it is meant to represent a dimensioned quantity then, without its associated dimension, it is meaningless. Students so often leave out the dimensions and lose marks for their ambiguity.

Here is a conversation I often have in class with my students to illustrate why that is so.

> Me Do you have any money on you?
> Him (*suspiciously*) Yes.
> Me OK, get out your wallet and lay a note on the table. (*He does so; it is a £10 note*) Feeling rich eh? So, here is a deal I can offer you. I'll give you two of my tens if you will give me your one ten. Agreed?
> Him (*greed overcoming suspicion*) Yes.
> Me OK, here are your twenty ... pence.

... and ethics prevent me from enforcing the deal; but amazingly it works every time!

One of the most catastrophic errors due to mistaken dimensions occurred with a 1998 space mission to Mars. A navigation error was caused because software was programmed in imperial pounds force rather than metric newtons which were what were required. The consequence was that the Mars Climate Orbiter crashed into the planet.

In 1984, an aircraft on a flight from the USA to Venezuela was overloaded by 13 tonnes because a shipper's weight which was in kilograms was assumed to be in pounds. In the previous year, a passenger airliner in Canada ran out of fuel in mid-flight because the fuel requirement was calculated in kilograms but loaded in pounds; less than half of what was needed.

When you are working out a problem you should consider what the right dimensions of the answer must be. If you evaluate a quantity by putting numbers into an equation don't forget that you are doing the same with the dimensions. For example, if you want to find the area of a circle which has a radius of 2 metres then perhaps you would use the formula for the area of a circle, which is $A = \pi r^2$, and then replace the r with 2. More correct would be to replace the r with 2 metres so that answer is $A = \pi\,(2\text{ metres})^2$ which leads to an answer of $A = 4\pi$ metres^2.

To find the dimensions of a quantity, simply put the dimensions into the formula. If you want to find the dimensions of force and you know the formula $F = ma$, where F is the force, m is mass and a is acceleration, then the dimensions of force are the product of the dimensions of mass and the dimensions of acceleration; kg × ms^{-2}. The density of an object is its mass divided by its volume, so the dimensions of density are kg ÷ m^3 which is kg m^{-3}.

Sometimes it is first necessary to *transpose formulae*, that is to rearrange the formula to make the subject a different variable. So, for example, if you want to find the dimensions of the universal gravitational constant, G (not to be confused with the acceleration due to gravity on the earth's surface, g), then start with the formula that gives the force of attraction, F, between two bodies of mass m_1 and m_2 which are at a distance r from each other, $F = Gm_1m_2r^{-2}$. Transpose this to get

$$G = Fm_1^{-1}m_2^{-1}r^2$$

and replace the variables with their dimensions.

Therefore, the dimensions of G are $\text{kg m s}^{-2}\ \text{kg}^{-1}\ \text{kg}^{-1}\ \text{m}^2$ which is $\text{kg}^{-1}\ \text{m}^3\ \text{s}^{-2}$.

Something to think about

1. Stress is force per unit area. If force is in kg m s^{-2} and area is in m^2, can you find the units for stress?

2. The kinetic energy of a mass, the energy it possesses by virtue of its motion, is given by $K = \frac{1}{2}mv^2$, where m is the mass and v is its velocity. Can you find the units of kinetic energy in terms of kilograms, metres and seconds?
3. Young's modulus of elasticity of a material, E, is the ratio of the tensile stress to the tensile strain and is given by the formula

$$E = \frac{FL}{A\delta L}$$

where F is the force applied (kg m s^{-2})
L is the length of the sample (m)
A is the cross-sectional area (m^2)
δL is the change in length (m)
Can you find the units of Young's modulus of elasticity?
4. The Stefan–Boltzmann constant, σ, is a constant of proportionality that relates the energy radiated by a black body to the body's surface area, temperature and time. It is given by:

$$\sigma = 7.5^{-1}\pi^5 k_B^4 h^{-3} c^{-2}$$

where k_B is the Boltzmann constant (J K^{-1})
h is the Planck constant (J s)
c is the speed of light (m s^{-1})
What are the units of the Stefan–Boltzmann constant?

1. kg m^{-1} s^{-2}
2. kg m^2 s^{-2}
3. kg m^{-1} s^{-2}
4. $(J\,K^{-1})^4\ (J\,s)^{-3}\ (m\,s^{-1})^{-2} = J\,K^{-4}\ m^{-2}\ s^{-1}$ which is Joules per degree Kelvin to the power of 4 per square metre per second.

Summary of the main points

- Keeping your work neat, tidy and ordered is very important. Try to keep your working aligned, one bit above another, so that it is very clear what is happening as you go from one line to the next. Be aware of the position and size of each number and symbol so that, for example, there is no possibility of mistaking ordinary numbers for superscripts or subscripts.
- We all make mistakes when working so, when you do so, ensure corrections are clear and unambiguous with a single line through each error and the correct value clearly written above, below or alongside.
- If you want to display data graphically, choose a form that does not hide or distort its true meaning. Avoid, where possible, axes that do not cross at the origin (i.e., where $x = y = 0$) unless there is good reason to do so. Steer clear of graphs that use complex or distorted views. Try not to show unrelated quantities, for example, absolute values and percentages, on the same graph.
- In experiments and fieldwork we are often called to draw conclusions from data. Be sure that the conclusions you draw are truly supported by the data. Predictions outside the range of the data are often suspect; consider the reliability of weather forecasts even just a week ahead.
- Different people write in different ways. Misunderstanding may occur because of the quality of your handwriting and that is important, but the way in which different people represent numbers, letters and symbols may also cause confusion.
- Do not be tempted to select experimental data that suits the conclusions you wish to draw. The outcome of your experiment might be leading you to an unexpected and therefore more interesting truth.
- ... and finally, numbers are just numbers. In the real world, numbers without their associated dimensions are all but meaningless.

4 Juggling with numbers

What do big and small numbers really mean?

Very large and very small numbers can be difficult to comprehend. If we talk of someone's salary as being in 'telephone numbers' we mean that he or she is paid a great deal more than we are getting. What does a million pounds mean to you? If you already have a million pounds then it is nothing exceptional. On the other hand, if all your worldly wealth amounts to no more than the small change in your pocket then a million pounds is beyond imagination. But what does a million pounds look like?

If you laid out one hundred pound coins in a row, that row would be 2.25 metres long. Now make a square of pound coins 2.25 metres by 2.25 metres, equivalent to the floor area of a small room; that would be £10 000. If each coin was the base of a pile of 100 pound coins, which would be 0.315 metres or just over one foot high, you would now have your million pounds laid out before you. See the picture on the next page. Not that impressive when you look at it this way is it? Even less so if viewed as a pile of banknotes? However, a million pounds in one pound coins weighs just less than ten tonnes, so don't try stacking them under your bed in an upstairs room.

The value of gold varies, but at the time of writing a million pounds worth weighs about 45 kg, or about half the weight of a well-built man.

So a million pounds can be visualised in a number of different ways. It could be a:

One-dimensional line of one pound coins, 22.5 kilometres long.
Two-dimensional square of one pound coins with side lengths of 22.5 metres.
Three-dimensional cube of one pound coins with side lengths of 1.17 metres.

One million pounds as a stack of £1 coins: the author dreams ...

Which is easiest to visualise? You could not see the end coins in a 22.5 kilometre line, even standing in the middle. A 22.5 metre square would require a very large room to lay it out, but you could touch the opposite sides of a 1.17 metre cube of pound coins. If large numbers can be modelled by two-, or better, three-dimensional shapes they can more easily be visualised.

Large quantities can also be easier to manage if the dimensions are chosen so that the numerical coefficient is within your range of imagination. You can probably imagine one, two, three, four or five of any quantity. For example, think of three one-pound coins.

You may see them as a line of three coins or as three coins touching in a triangle or as coins, one on top of the other. You can probably count them in your hand without looking at them. However, this does not work so well if you have 10 coins, or 30 coins or 100 coins or 300 coins or 1000 coins and so on. The more there are; the harder they are to visualise.

Imagine the distance from London to Penzance in the south-west corner of England. If we describe it as half a million metres, that description is unmanageable; although a metre is an easy length to visualise, half a million is a very large number; 300 miles might be better. A mile is the distance walked in about 20 minutes and 300 is a smaller although still a somewhat large number. If you drive at an average of 50 miles per hour then the distance may be represented by its journey time of 6 hours, quite a small number of hours. If you allow for breaks, then that is approximately a journey of one day. One is a very easy number to visualise.

Curiously, time is a very flexible, if invisible, dimension because we can appreciate a wide range of time periods:

A second, a slow heart beat.
A minute, the time to walk the length of a football pitch.
An hour, an average lecture.
A day, from breakfast to the next breakfast.
A week, one Saturday night out to the next.
A month, one credit card payment to the next.
A year, one birthday to the next.

You may think of any of these periods in different but more personal terms.

Time is a useful way of dealing with enormous distances. Light travels at about 300 000 kilometres per second. Light-time is the distance that light travels in a given time, so a light second is 300 000 km. That means that the distance from:

Belfast to Istanbul, 3000 km, is one hundredth of a light-second.

The earth to the moon is 1.28 light-seconds.
The earth to the sun is 8.33 light-minutes.
The earth to Proxima Centauri, our nearest star, is 4.24 light-years.
The earth to the edge of the visible universe is about 14 000 000 000 light-years.

This last distance would be 130 000 000 000 000 000 000 000 kilometres, a staggering and hence meaningless number.

The kilogram as a unit of mass is easy to recognise; it is the weight of a litre of water. Your body weighs several tens of kilograms. A car weighs about a tonne. A large truck may be 44 tonnes. The ocean liner, the QE2 weighs 70 000 tonnes. Mount Everest is approximately 3 000 000 000 000 tonnes. The earth is 6 000 000 000 000 000 000 000 tonnes. The mass of the sun is 330 000 times the mass of the earth.

In the case of mass, length and time, the quantities that are easiest to visualise are those which are relatively small multiples of units that are within our range of everyday experience. They would be distances from one millimetre through one metre to one kilometre, masses from one gram through one kilogram to one tonne and time from one second through one day to one year.

So how can we relate large and small quantities? Standard form is one useful way of comparing the relative magnitudes of apparently incompatible quantities. A number that is written in standard form places greater emphasis on the number's order of magnitude than its absolute value. So the masses of the objects described above, in standard form, would be:

a litre of water	1.0 kg,
a car	1.0×10^3 kg,
the QE2	7.0×10^7 kg,
Mt Everest	3.0×10^{15} kg,
the earth	6.0×10^{24} kg and
the sun	2.0×10^{30} kg.

Notice that it is the power of 10 that is giving most of the information about the relative masses of these bodies. If you ignore the mantissa, the number before the × sign, these are the orders of magnitude greater than a litre of water: a car is 3, the QE2 is 7, Mt Everest is 15, the earth is 24 and the sun is 30. These numbers are much more manageable.

Of course the same applies to small numbers. The mass of a raindrop is about 0.00001 kg which is 1.0×10^{-5} kg. A molecule of water is 3.0×10^{-23} kg. An electron is 9.0×10^{-31} kg. So these are the number of orders of magnitude less than a litre of water: a raindrop is 5, a molecule of water is 23, and an electron is 31.

In computer storage, we talk in terms of bytes, kilobytes, megabytes, gigabytes and terabytes; each one is 3 orders of magnitude greater than the previous one. So 1 KB is 10^3 bytes, 1 MB is 10^6 bytes, 1 GB 10^9 bytes and 1 TB 10^{12} bytes. As technology progresses over the coming years, we will talk of petabytes (10^{15} bytes), exabytes (10^{18} bytes), zettabytes (10^{21} bytes), yottabytes (10^{24} bytes), etc.

Something to think about

1. How many raindrops does it take to fill a lake (assumed to be 10 km long by 2 km wide by 500 m deep)? One litre of water weighs 1 kilogram and occupies 0.001 m^3.
2. Could the population of the UK stand, holding hands, around the coast of Africa?
3. In the Lewis Carroll poem …

> The Walrus and the Carpenter were walking close at hand;
> They wept like anything to see such quantities of sand:
> 'If this were only cleared away,' they said, 'it would be grand!'
> 'If seven maids with seven mops swept it for half a year.
> Do you suppose,' the Walrus said, 'that they could get it clear?'
> 'I doubt it,' said the Carpenter, and shed a bitter tear.

So was the Carpenter right to be doubtful?

1. 1 000 000 000 000 000 000 = 10^{18} raindrops.
2. Yes, but they would be shoulder to shoulder. This unlikely result is because the 61 million or so who live on these small islands are spread across its whole area, but if stretched in a long line around the 32 000 kilometres of Africa's mainland coastline would have just half a metre each to stand in.
3. That rather depends on the productivity of each maid and how much sand there was. So how much sand is there? One estimate on the internet is that there are 7.5 × 10^{18} grains of sand on earth. So since the world's coastline is about 356 000 km long and assuming that half of that is sand, then there are 4 × 10^{10} grains per metre of beach shoreline and it is doubtful that the seven maids, even in a full lifetime, could shift that.

The significance of figures

How many figures?

When you have done a sum, how precisely should you quote the answer?

Imagine that you have just measured the radius of a bicycle wheel and found it to be 45 centimetres and then used the formula $c = 2\pi r$ to find its circumference. With your calculator set to display in 'scientific' mode, and using the value of π in your calculator, the answer displayed by the calculator is 282.743339 centimetres. How precise is that? The precision of the computation is good to nine significant figures but is that the precision of the measure of your circumference? If it was, it would imply that you have calculated the circumference to one millionth part of a centimetre, based upon a measurement that was good to the nearest centimetre.

A good rule of thumb is that results should be quoted to no more than one more significant figure greater than the number of significant figures associated with the least significant contributing statistic. In this case, the radius was measured to two significant figures, so the circumference should not be quoted to more than three significant figures, 283 centimetres. In this case, we are still claiming a precision greater than the original measurement, $\frac{1}{283} < \frac{1}{45}$, but at least we are not introducing any substantial rounding errors. In this case, if we had given the answer to the same number of significant figures as the measurement of the radius, then the precision of the circumference would have been less than that of the original measurement $\frac{1}{45} < \frac{10}{280}$.

Trying to impress

A school in the USA recently reported that it had 92.84% of students at or above state standards. It also claimed to have 2463 students enrolled. Are these statistics compatible? The number of significant digits in the 92.84% statistic implies that the true value lies between 92.835% and 92.845% which in turn implies that the error in the 92.84% statistic is less than 0.005%; 0.005% of 2463 students is less than $\frac{1}{8}$of a student and 92.84% of 2463 students is 2286.6 students. How did they get 0.6 of a student to state standards? Therefore, the precision of such a claim is overstated and cannot be justified. A much more justifiable claim would have been for 92.8% of the students.

Statistical implication

Ivory Soap claims to be 99.44% pure. Does that imply that the purity, whatever that is, is so tightly controlled that it does not go below 99.435% or above 99.445%? Not in this case, because the '99.44% pure' is in fact a trademark not a statement of fact; but who would appreciate the difference? Is this an example of using statistics to give the appearance of scientific credibility?

Losing the plot

On the other hand, conclusions can be misleading by understating their precision. Here are the votes for candidates in an election:

Alfred Allen	2345
Betty Brown	6202
Charlie Cooper	6178

In reporting the results of the election, the percentages of the votes were shown as Allen 16%, Brown 42% and Cooper 42%. These percentages have been correctly calculated to the nearest 1% and they do add up to 100%. However, an essential feature of an election is that there is only one winner, unless the two top candidates gain exactly the same number of votes. So although it is true that Brown and Cooper were very close, it was Brown who won. It would therefore have been better if the reporting showed the results as Brown 42.1%, Cooper 42.0% and Allen 15.9%.

Something to think about

1. A bag of fertiliser is sufficient for 8.7 m^2 of ground. How much ground are 233 bags sufficient for?
2. If your weekly pay increases from £387 to £396 what percentage rise is this?
3. Three candidates in an election poll 9849, 9856 and 9864 votes, respectively. In quoting the results in percentage terms, what would be the most sensible number of significant figures to use?

1. $8.7 \times 233 = 2027.1$, but since the coverage of a bag is quoted to two significant figures then the coverage of the area should not be quoted to more than three significant figures. A sensible answer to the problem would be 2030 m^2.

2. This is a rise of £9 which is $\frac{9}{387} \times 100\% = 2.32558\%$. However, the rise in pounds is only calculated to one significant figure so the rise in percentage terms should only be quoted to two significant figures. This is therefore a rise of 2.3%.
3. The votes, to six significant figures, are respectively 33.3085%, 33.3322% and 33.3593%. Of course, six significant figures are meaningless as the number of votes is only of four figures. If the votes were shown to two significant figures that would give 33%, 33% and 33% which would suggest a three-way tie. To three significant figures the votes are 33.3%, 33.3% and 33.4% which does indicate one winner, although it still suggests a tie for second place. If that is still considered misleading then four significant figures would be required and the votes would need to be reported as 33.31%, 33.33% and 33.36%.

Errors in measurement

Whenever you do a science or an engineering practical workshop, laboratory exercise or fieldwork investigation you usually get it wrong. No, that is not meant as criticism, but it is invariably true because you make measurements and you take observations.

Simply put, it is impossible to make exact measurements for several good reasons. If the quantity that you are trying to measure is continuous, which most quantities are, then measurements are limited by the resolution of the measuring device. If you measure your own height at home, for example, how would you do it? You might stand against a wall, facing it, place a book on your head and mark a line on the wall and then measure from floor to the mark on the wall with a tape measure. So what could possibly go wrong? Quite a lot actually; here are some possible sources of error:

Do you take account of the thickness of your shoes or socks?
How much do you stoop when holding a book above your head?
Is the book horizontal?
How soft is the carpet?
Is the tape damaged?
What is the least count of the tape, centimetres or millimetres?

Your footwear and how you stand are two examples of factors that affect the quantity that is to be measured. Wearing shoes will produce a significant bias if the measured height is assumed to be barefoot. Holding the book horizontal, using a damaged tape and unquantified sinking into the carpet will affect the measuring process. Deciding the location of the exact point at the bottom of the wall, and the pencil line above, determine what is actually measured. The resolution of the tape limits the accuracy of the quantity measured.

All of these factors create error. As the measurer, you, hopefully, take precautions to ensure that the errors are managed and minimised. Although you will produce a measurement at the end of the exercise, that measurement will contain uncertainty. Given the above factors, it is doubtful that you could measure your own height even to the nearest centimetre.

But these are not the only types of error that you may make. We all occasionally make big mistakes; blunders, we might call them. The best advice on making blunders is that, if you cannot eliminate them from your work, then make big ones. Big mistakes are much easier to identify than small ones. Better of course is to have a method of identifying the blunders when they are made so that they can be eliminated from your experimental data.

For example, if you want to measure a distance between one place and another, measure it in both directions, get two different people to measure it, use two different tapes, measure it in different units such as feet and metres and then convert both measurements to the same units. All of these mean that the second measurement is an almost independent check upon the first one.

You can improve your estimate of a given quantity by making many repeated measurements and then taking the average of them. This will improve the precision of your estimate but will not necessarily improve the accuracy much. Precision is a measure of repetition whereas accuracy is the difference from the truth. So, for example, if you measure a distance in yards but assume you are measuring in metres, then although the standard deviation of the mean of the set of observations will become smaller, that is improved with more measurements, the average answer will still be about 10% in error.

Analogue measuring instruments are particularly vulnerable to parallax measurement errors. Any device that has an indicator that is not in contact with the measurement scale needs to be read with the scale perpendicular to the observer's line of sight. By moving your head from side to side you can make the apparent measurement change. So, take care not only with clocks and watches of course but also with mercury or alcohol thermometers, barometers, micrometers and voltmeters, in fact anything with a hand and a dial. View it square on to avoid parallax error.

So how can you estimate the uncertainty of a quantity derived from independent and unbiased measurements? For example, you wish to estimate the volume of a rectangular swimming pool but can only measure each dimension with an estimated uncertainty. What will be the uncertainty of the calculated volume?

If a variable, v, is a function of three unrelated variables, l, b, and d, then the standard deviation of v, σ_v, is given by:

$$\sigma_v^2 = \left(\frac{dv}{dl}\right)^2 \sigma_l^2 + \left(\frac{dv}{db}\right)^2 \sigma_b^2 + \left(\frac{dv}{dd}\right)^2 \sigma_d^2$$

That is, it is the sum of the products of the squares of the partial differentials of the function with respect to each variable and the square of the standard deviation of the variable.

So if the volume, v, is the product of length, breadth and depth, l, b and d, that is $v = lbd$, and the standard deviations of l, b and d, are σ_l, σ_b and σ_d, then:

$$\sigma_v^2 = (bd)^2 \sigma_l^2 + (ld)^2 \sigma_b^2 + (lb)^2 \sigma_d^2$$

If l, b, and d are measured as 50 m, 10 m and 2 m, respectively and the standard deviations of the measurements are all 0.1 m:

$$\sigma_v^2 = (10 \times 2)^2 0.1^2 + (50 \times 2)^2 0.1^2 + (50 \times 10)^2 0.1^2 = 2604 m^6$$

$$\sigma_v = 51 m^3$$

Something to think about

1. You wish to estimate the speed of your model aeroplane, so you decide to time its flight over a distance marked out on the ground. You get two friends to stand at the ends of the marked distance and fly the aircraft in a straight line over their heads. You instruct your friends to shout when the aircraft flies overhead. What do you think will be the sources of error in your calculation of the aircraft's speed?
2. You need to fill your bath but cannot be bothered to sit and wait until it is full. You decide to measure the length, width and depth of the bath which are 1.5 m, 0.7 m and 0.3 m, respectively. You also find that you fill a 1 litre drinks bottle from the tap in 2.5 s, and from that you estimate the flow rate from the tap. Since 1 m^3 is 1000 litres, that gives a flow rate of 0.0004 m^3s^{-1}. However, your bath, like most, is not rectangular so you have some uncertainty in your dimensions and estimate the standard deviations to be 0.1 m for length and breadth, 0.05 m for depth and 0.000 05 m^3s^{-1} for flow rate. How long will it take to fill the bath and with what uncertainty of time?

1. Defining the end points.
   Measuring between the end points.
   Flying the aircraft horizontally.
   Flying the aircraft in a straight line.
   Making the straight line pass over the end points.
   The speed of the wind over the ground along the line of flight.
   The speed of the wind over the ground at right angles to the line of flight.
   The ability of the friends to identify the instant that the aircraft is above them.
   The time it takes for them to react and shout.
   Your time of reaction from hearing a shout to pressing the stop watch.
   Your ability to get the sums right.
2. The volume of the bath is $v = lbd$, where length, breadth and depth are l, b and d, respectively. If the flow rate is f then the time, t, taken to fill the bath is $t = lbdf^{-1}$. So:

   $$t = 1.5 \times 0.7 \times 0.3 \div 0.0004 = 787.5\,\text{s}$$

   which is 13 minutes and 7.5 seconds.

   The uncertainty of t is given by:

   $$\sigma_t^2 = \left(\frac{dt}{dl}\right)^2 \sigma_l^2 + \left(\frac{dt}{db}\right)^2 \sigma_b^2 + \left(\frac{dt}{dd}\right)^2 \sigma_d^2 + \left(\frac{dt}{df}\right)^2 \sigma_f^2$$

   $$\sigma_t^2 = \left(bdf^{-1}\right)^2 \sigma_l^2 + \left(ldf^{-1}\right)^2 \sigma_b^2 + \left(lbf^{-1}\right)^2 \sigma_d^2 + \left(-lbdf^{-2}\right)^2 \sigma_f^2$$

$$\sigma_t^2 = \left(\frac{0.7 \times 0.3}{0.0004}\right)^2 0.1^2 + \left(\frac{1.5 \times 0.3}{0.0004}\right)^2 0.1^2$$

$$+\left(\frac{1.5 \times 0.7}{0.0004}\right)^2 0.05^2 + \left(-\frac{1.5 \times 0.7 \times 0.3}{0.0004^2}\right)^2 0.00005^2$$

$\sigma_t^2 = 42329 \text{ s}^2$

$\sigma_t = 206$ s, which is 3 minutes and 26 seconds.

So it is worth checking on the bath after $10(=13-3)$ minutes to make sure it does not overfill.

Interest rates—calculation of AER and APR

Interest rates

Lending money is big business. Borrowing money comes with a price. If you want to borrow money then you will normally have to pay interest for the privilege. The amount of interest you pay will depend upon the terms and conditions of the loan. In practice, what you pay will depend on the principal, that is, the amount you borrow, and the interest rate. Interest is usually 'compounded', that is, each time it is due to be paid, it is added to the principal if it is not paid off promptly.

Annual Equivalent Rate

Annual Equivalent Rate, AER, is the rate of interest when it is calculated once a year. So, for example, if the principal is £1000 and the interest rate is 5%, then after 1 year the amount owing will be £1050. If the interest is not repaid as it accrues then after 10 years the amount owing will be £1628.89 because interest due becomes payable on any previously unpaid interest. This simple table shows the amount of interest accrued assuming that interest is not paid off. It shows how a high interest rate can easily make a debt unmanageable.

Principal £1000	*Interest added to principal*			
Interest rate	*1%*	*5%*	*10%*	*20%*
1 year	£10.00	£50.00	£100.00	£200.00
5 years	£51.01	£276.28	£610.51	£1488.32
10 years	£104.62	£628.89	£1593.74	£5191.74
20 years	£220.19	£1653.30	£5727.50	£37337.60

Annual Percentage Rate

Annual Percentage Rate, APR, is the rate of interest calculated on the basis that interest is calculated periodically and added at the end of each period in the year. For example, if the period is 1 month and the periodic rate is 2%, then the APR is 24%. However, if interest is added at 2% per month, that is compounded on a monthly basis, then the AER will be 26.8% ($=[1.02^{12}-1]\times 100\%$). At low interest rates the difference between AER and APR compounded monthly is very small. However, as interest rates become more extortionate, the true AER may be disguised by the APR if compounded monthly. The table below shows the AER equivalent to a given APR compounded monthly and compounded daily.

APR (%)	*Equivalent AER for APR compounded monthly (%)*	*Equivalent AER for APR compounded daily (%)*
1	1.00	1.01
5	5.12	5.13
10	10.47	10.52
20	21.94	22.13
50	63.21	64.82
75	106.99	111.54

Interest rates as large as 75% and greater may be applied to hire purchase agreements for some electrical and other retail goods. By signing an agreement like this you may end up paying many times the principal, that is, retail value of the goods, in interest. A recent television advertisement offers loans at a staggering rate of 2689%.

Something to think about

1. You are interested in buying a car from a dealer. The price of the car is £10 000 but you cannot afford this. The dealer offers you a loan with an interest rate of 2% per month with a monthly payment from the end of the first month of £249.11 per month. You think that this sounds like a good deal; one that you can afford and so you sign the forms. Having driven home you decide to work out how long it will take to pay the loan off and how much interest you will have to pay on top of the original loan.
2. You might have been surprised by the outcome in 1. above. Now consider the repayment time and the total of the interest to be paid if the monthly repayment was reduced to £220.48.
3. What would be the repayment time and the total of the interest to be paid if the monthly repayment was increased to £287.68?

1. 6 years and 10 months, £10 427 interest. So it will take almost 7 years and you will have paid more in interest than the original cost of the car. There is no simple formula for this calculation but there is a simple process that you can use. To find what you owe at the end of the first month add the interest to the initial amount owing, 2% of £10 000, and subtract the first payment:

$$£10\,000 + 0.02 \times £10\,000 - £249.11 = £9950.89.$$

To find what you owe at the end of the next month add the interest to the amount owing, 2% of £9950.89, and subtract the second payment. Repeat for each month until all the loan is paid off. The next few lines of the calculation are:

£9950.89 + 0.02 × £9950.89 − £249.11 = £9900.80
£9900.80 + 0.02 × £9900.80 − £249.11 = £9849.70
£9849.70 + 0.02 × £9849.70 − £249.11 = £9797.59

Notice that although you pay £249.11 pounds a month, you are only reducing your debt by about £50 per month at the start. The easiest way to do this calculation is with a spreadsheet.

2. 10 years, £16 458 interest.
3. 5 years, £7 261 interest.

What sort of average?

The words *average* and *mean* are often used to denote the same thing but in fact they have quite different meanings in mathematics. Average is a measure of *central tendency*. One such measure is the mean. Central tendency is usually described by the arithmetic mean, the mode or the median. The measure of central tendency that is most useful depends upon the data set being investigated and how you intend to use that central tendency statistic.

The arithmetic mean is the measure of central tendency you are probably most familiar with. It is found by taking all the elements of the data set, adding them up, and then dividing by the count of them. For example, you wish to get the best estimate that you can for the length of a room so you decide to get many independent measurements. You persuade nine different people, each with a different tape, to make the measurement. Here are the results, all in metres:

$$4.769, 4.803, 4.764, 4.768, 4.738, 4.783, 4.936, 4.722, 4.791$$

So the average is:

$$\frac{4.769 + 4.803 + 4.764 + 4.768 + 4.738 + 4.783 + 4.936 + 4.722 + 4.791}{9} = 4.786\,\text{m}$$

The average is the most useful indicator of central tendency when the measurements are at least theoretically continuous, that is they can take any value unlimited by places of decimal, and all are not biased by anything other than the random errors of measurement. Most experimental and field data should fall into this category.

But what if you suspect that your data may not be reliable but do not have the time to inspect them all? It might be that one or more of your measurements has a large error. If that is so, then the arithmetic mean will also have a large error. One way around this problem is to find the median value.

The median is simply the middle value. All you have to do is take your measurements and then rank them in order, from smallest to largest, and then identify the one in the middle; the fifth one in this example:

$$4.722, 4.738, 4.764, 4.768, \underline{4.769}, 4.783, 4.791, 4.803, 4.936$$

So the median value is 4.769 m. By examining this string of data it is clear that the last number is significantly greater than the rest and is therefore probably the result of an error of measurement. Without this measurement the arithmetic mean is 4.767 m which is very close to the median value. The median value can be used whether the data is continuous or discrete.

The mode lends itself best to discrete data since it is the value of a set that appears most often. If the data is continuous then, theoretically, no value will appear more than once. The mode is the value that is found by a majority vote process. For example, a number of managers wish to select the optimum size of a team for a particular task. Here are their opinions ranked in order:

2, 2, 4, 4, 4, 4, 4, 4, 5, 5, 6, 6, 6, 7, 7, 8, 10, 11

To take the arithmetic mean, 5.5, would itself be meaningless; you cannot have half a person in a team. The median value of 5 is one of the less popular choices but the mode, 4, represents the most popular team size.

Something to think about

1. What are the arithmetic mean, median and mode of this data set?

   32, 36, 32, 36, 33, 28, 28, 34, 35, 32, 33, 34

2. What are the arithmetic mean, mode and median of this data set?

   5.5, 5.9, 5.6, 5.8, 5.3, 5.4, 5.4, 5.5, 5.4, 5.6, 5.7, 5.1, 5.6, 5.2, 5.9, 5.5, 5.4, 5.4, 5.3, 6.3

3. Bags of cement are being filled at a cement works where the production line is running smoothly. As part of the quality control process bags to be weighed are selected at random. Which is the best measure of the average weight of a bag; the arithmetic mean, the mode or the median of the set of weights?
4. At the same cement works on another production line the filling mechanism occasionally sticks so that some bags are significantly overfilled. Bags are later selected at random. Which is the best measure of the average weight of a bag; the arithmetic mean, mode or median of the set of weights?

5. In the cement works there are a number of minibuses for moving workers about the site. How would you find the capacity of the average bus?
6. In the car park at the cement works there are many cars and each car is one of only 20 distinctive colours. How would you find the average colour?
7. If there is an even number of members in a data set and all members have a different value, how would you choose the median value?

1. Mean 32.75, median 33, mode 32.
2. Mean 5.43, mode 5.4, median 5.5.
3. Arithmetic mean, because the weight measurements are theoretically continuous and there are not expected to be any biased measurements.
4. Although the weight measurements are theoretically continuous there are a number which are known to be biased; therefore the median value is most appropriate.
5. The average bus is the one of which there are most representatives of its type, so the mode of the bus capacities would be the best average.
6. The average colour is the colour that appears most often and as such must be the mode of the set. Notice that with the mode the data set does not even need to be numerical. The arithmetic mean or median would be meaningless in this context.
7. With an even number of members there is no single central member, so take the two central members and find their arithmetic mean.

What mean do you mean?

The arithmetic mean is not the only mean; there are several others that appear in mathematics. Here are some of them.

The *weighted mean* is similar to the arithmetic mean but takes account of the different importance of each member of the set. In the room-measuring example of the previous section it may be that some of the measurements are more trustworthy than others. It could be that one person was known to be inexperienced or was using an old tape of lesser quality. If that is so, then a weight is applied to each measurement. The equation for the weighted mean is:

$$\frac{\sum_{i=1}^{i=n} w_i x_i}{\sum_{i=1}^{i=n} w_i}$$

where the members of the data set are $x_1, x_2, x_3, \ldots x_n$ and have weights $w_1, w_2, w_3, \ldots w_n$.

We now use the same observations as before but with weights of 1, 2 and 3 allocated to different measurements, based upon their assumed relative quality. That means that we take twice as much notice of some measurements because they are believed to be twice as good as the worst measurements, and three times as much notice of the best measurements, compared with the worst, because they are believed to be three times as good.

Measurement:	4.769,	4.803,	4.764,	4.768,	4.738,	4.783,	4.936,	4.722,	4.791
Weight:	1,	1,	2,	2,	2,	3,	3,	3,	3

The weighted mean is:

$$\frac{1 \times (4.769 + 4.803) + 2 \times (4.764 + 4.768 + 4.738) + 3 \times (4.783 + 4.936 + 4.722 + 4.791)}{2 \times 1 + 3 \times 2 + 4 \times 3}$$

$$= \frac{95.808}{20} = 4.790\,\text{m}$$

The *geometric mean* is found as the nth root of the product of the n members of the data set. If the members of the data set are $x_1, x_2, x_3, \ldots x_n$ then the geometric mean is:

$$\sqrt[n]{x_1 \times x_2 \times x_3 \times \ldots \times x_n} = \left(\prod_{i=1}^{i=n} x_i\right)^{\frac{1}{n}}$$

So the geometric mean of 2, 3, 9 and 24 is $\sqrt[4]{2 \times 3 \times 9 \times 24} = 6$. Likewise, the geometric mean of $\frac{1}{4}$, $\frac{1}{8}$, $\frac{1}{20}$, $\frac{1}{50}$ and $\frac{1}{100}$ is $\sqrt[5]{\frac{1}{4} \times \frac{1}{8} \times \frac{1}{20} \times \frac{1}{50} \times \frac{1}{100}} = \frac{1}{20}$.

If a and b are the lengths of the sides of a rectangle, then the geometric mean of a and b is the length of the side of a square that has the same area as the rectangle. Similarly, if a, b and c are the sides of a right rectangular prism, then the geometric mean of a, b and c is the length of the side of a cube that has the same volume as the right rectangular prism.

Geometric mean is particularly useful for some financial calculations. For example, the value of a business in millions of pounds in four successive years is assessed to be 1.00, 1.68, 1.71, 2.43 and 2.58. What is the average annual percentage increase in value?

If you calculate the percentage increase for each successive year, that comes out to be 68.0%, 1.8%, 42.1% and 6.2%. The arithmetic mean of these values is 29.5%. If you increase 1.00 by 29.5% four times you get 2.81. This is not the 2.58 you were seeking.

However, if you convert the percentage increases to the ratio equivalents, that is, divide by 100 and then add 1, the successive ratios are 1.680, 1.018, 1.421 and 1.062. Now find the geometric mean of these values:

$$\sqrt[4]{1.680 \times 1.018 \times 1.421 \times 1.062} = 1.267$$

Convert that ratio back to a percentage; 26.7%. Now increase 1.00 by 26.7% four times and you get 2.58. So to find a mean percentage increase it is essential to use a geometric mean.

The *harmonic mean* is the inverse of the means of the inverses of the members of the set. So if the members of the data set are $x_1, x_2, x_3, \ldots x_n$ then the harmonic mean is:

$$\frac{1}{\left(\frac{\frac{1}{x_1}+\frac{1}{x_2}+\frac{1}{x_3}+\ldots+\frac{1}{x_n}}{n}\right)} = n\left(x_1^{-1}+x_2^{-1}+x_3^{-1}+\ldots+x_n^{-1}\right)^{-1}$$

$$= n\left(\sum_{i=1}^{i=n} x_i^{-1}\right)^{-1}$$

Suppose you calculate the speed of a car which is going around a track that is 10 km long, and find the following values, all in km/hr:

180, 150, 90, 125 and 160

What is the average speed? It would be tempting to find the arithmetic mean of those speeds as 141 km/hr but this would be incorrect. With these speeds the times around the circuits are 200, 240, 400, 288 and 225 seconds, respectively. The total time is therefore 1353 seconds. As the total distance is 50 km then the average speed is the total distance divided by the total time, which is:

$$\frac{50 \times 3600}{1353} = 133\, km/hr$$

And this is the harmonic mean:

$$5 \times \left(180^{-1}+150^{-1}+90^{-1}+125^{-1}+160^{-1}\right)^{-1} = 133$$

The quadratic mean or *root mean square* (RMS) makes it possible to find a meaningful 'average' value when the arithmetic mean is not helpful.

To create the RMS of a data set:

Square each of the members of the set
Add up the squares
Divide by the count of the members
Find the square root of this value.

This is the same as using the formula:

$$\sqrt{\frac{x_1^2+x_2^2+x_3^2+\ldots+x_n^2}{n}}=\sqrt{\frac{1}{n}\sum_{i=1}^{i=n}x_i^2}$$

The RMS of a continuous function may be found by integration from:

$$\sqrt{\frac{1}{b-a}\int_a^b [f(x)]^2\,dx}$$

where a and b are the limits of integration and $f(x)$ is the function.

For example, the arithmetic mean of a sine wave plotted over integer multiples of a whole cycle is always zero irrespective of the magnitude of the sine wave. This is because the area below the x-axis is the same shape and size, a mirror image, of the area which is above. However, the RMS is proportional to the magnitude of the sine wave. So if $f(x)=A\sin x$ over the limits 0 to 2π then:

Arithmetic mean $=\frac{A}{2\pi}\int\limits_0^{2\pi}\sin x\,dx=0$, but

$$\text{RMS}=\sqrt{\frac{A^2}{2\pi}\int_0^{2\pi}\sin^2 x\,dx}$$

$$=\sqrt{\frac{A^2}{2\pi}\int_0^{2\pi}\frac{1}{2}(1-\cos 2x)\,dx}$$

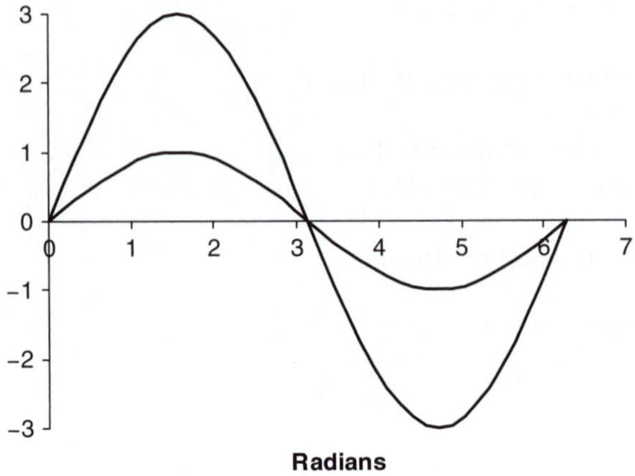

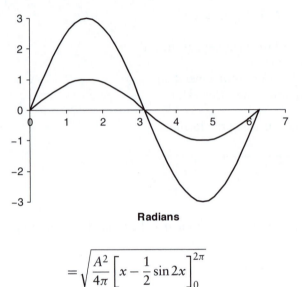

$$= \sqrt{\frac{A^2}{4\pi}\left[x - \frac{1}{2}\sin 2x\right]_0^{2\pi}}$$

$$= \sqrt{\frac{A^2}{4\pi}2\pi} = \frac{A}{\sqrt{2}}$$

Just taking the simple numbers 1, 2, 3, 4 and 5:

RMS $\sqrt{\frac{1^2+2^2+3^2+4^2+5^2}{5}} = 3.32$

Arithmetic mean $\frac{1+2+3+4+5}{5} = 3.00$

Geometric mean $\sqrt[5]{1\times 2\times 3\times 4\times 5} = 2.61$

Harmonic mean $5\times\left(1^{-1}+2^{-1}+3^{-1}+4^{-1}+5^{-1}\right)^{-1} = 2.19$

Notice the order, which is true for any data set:

RMS $\geq$ Arithmetic mean $\geq$ Geometric mean $\geq$ Harmonic mean

Something to think about

1. What are the RMS, arithmetic mean, geometric mean and harmonic mean of this data set: 3, 5, 7 and 9?

In the following questions which mean would be the most useful to find?

2. The mean speed of an aircraft given the speeds of several flights between the same pair of airports.
3. The mean time of flight of an aircraft given the times of several flights between the same pair of airports.
4. The mean deviation in altitude of an aircraft where the pilot is trying to fly at a constant altitude.
5. The ratio of improvement in journey times by rail from London to Edinburgh, measured every 10 years.

1. RMS 6.40, arithmetic mean 6.00, geometric mean 5.54, harmonic mean 5.08.
2. Harmonic mean.
3. Arithmetic mean.
4. RMS.
5. Geometric mean.

Ratios of ratios

Your old bed is getting lumpy and you decide to buy a new one. In the paper you find an advertisement that says that in the New Year sale there is 40% off on all beds, even those that have already been discounted by 30%. In the last week of the sale there will be an additional 20% and on Saturday, because it is the store owner's birthday, he will be giving an extra 10% on top of everything else. You add up all the discounts on offer and conclude that 40%, 30%, 20% and 10% added together makes 100% off, and so you will get the £1000 bed you had your eye on for free. Is this a really good deal?

Of course it is not. As the saying goes, if it looks too good to be true, it probably is. Discounts are applied successively, not lumped together as one. In other words, it is not discounts that are added but ratios that are multiplied together. So a 40% discount means that you pay 0.6 of the previous price. A 30% discount gives a ratio of 0.7 and the 20% and 10% discounts give ratios of 0.8 and 0.9, respectively.

So with all these discounts applied to the bed, initially advertised as £1000, that means it will cost:

$$\pounds 1000 \times 0.6 \times 0.7 \times 0.8 \times 0.9 = \pounds 302.40$$

So this is less than 70% off what was probably originally a rather pricey item.

On a bicycle, the number of revolutions of the back driving wheel to each full rotation of the pedal is the ratio of the number of teeth on the circle of the pedal to the number of teeth on the gear circle of the wheel. So if the pedal has 35 teeth and the circles on the rear wheel have 15, 17, 20, 23 and 27 teeth then the gear ratios are $\frac{35}{15} = 2.33$, $\frac{35}{17} = 2.06$, $\frac{35}{20} = 1.75$, $\frac{35}{23} = 1.52$ and $\frac{35}{27} = 1.30$. However, for best comfort when cycling an equal change in these ratios is not what is ideally required but a constant ratio of the ratios. So, relative to the next successive ratio, the ratios of ratios are:

$$\frac{2.33}{2.06} = 1.13, \frac{2.06}{1.75} = 1.18, \frac{1.75}{1.52} = 1.15 \text{ and } \frac{1.52}{1.30} = 1.17$$

These represent 13%, 18%, 15% and 17% reductions in speed of the bicycle at each gear change for a constant pedal speed.

However, those ratios of gear ratios are also the ratios of the number of teeth on successive circles.

$$\frac{17}{15} = 1.13, \frac{20}{17} = 1.18, \frac{23}{20} = 1.15 \text{ and } \frac{27}{23} = 1.17$$

Something to think about

1. A 25% discount makes the original price 75% of what it was, but how many times does '25% off' have to be applied before the cost of an item is not more than 25% of its original cost?
2. How many times does '10% off' have to be applied until the cost of an item is not more than 10% of its original cost?
3. On a bicycle there are five circles on the back wheel with 15, 17, 20, 23 and 27 teeth and three circles on the pedal with 35, 46 and 59 teeth. How many useful gear ratios are there if a change of less than 6% is considered not to be useful?
4. In a construction company the colour of a hard hat usually shows the function of the wearer. However the hard hats have become mixed up so that $\frac{5}{7}$ of the managers now wear $\frac{3}{4}$ of the white hard hats. (The other $\frac{2}{7}$ of the managers wear coloured hard hats and the remaining $\frac{1}{4}$ of the white hard hats are now worn by non-managers). Each person has only one hard hat. What is the ratio of white hard hats to managers in this company?

1. '25% off' gives a ratio of 0.75. The problem can be reduced to that of finding n in the equation $0.75^n = 0.25$ so, on 'taking logarithms', $n = \frac{\log 0.25}{\log 0.75} = 4.8$ which, when rounded up, is 5. Therefore, a 25% discount must be applied five times to reduce the cost to not more than 25% of its original amount.
2. 10% of the original cost is 0.1. '10% off' gives a ratio of 0.9. The problem can be reduced to that of finding n in $0.9^n = 0.1$ so, on taking logarithms, this leads to $n = \frac{\log 0.1}{\log 0.9} = 21.9$. The 10% discount must be applied 22 times to reduce the cost to not more than 10% of its original amount.
3. With three circles on the pedal and five circles on the back wheel there are potentially 15 different gear ratios.

This table shows all combinations of pedal and wheel circles, and their associated ratios listed in order of increasing ratios. The third column shows the ratio of the ratio on the current line to the ratio on the previous line and the final column shows the same statistic as a percentage change. In the second table the pedal/wheel combinations that are not useful, those with a value of less than 6%, have been removed.

Teeth pedal/wheel	*Ratio*	*Ratio of ratio*	*Change (%)*
35/27	1.30		
35/23	1.52	1.17	17.4
46/27	1.70	1.12	12.0
35/20	1.75	1.03	2.7
46/23	2.00	1.14	14.3
35/17	2.06	1.03	2.9
59/27	2.19	1.06	6.1
46/20	2.30	1.05	5.3
35/15	2.33	1.01	1.4
59/23	2.57	1.10	9.9
46/17	2.71	1.05	5.5
59/20	2.95	1.09	9.0
46/15	3.07	1.04	4.0
59/17	3.47	1.13	13.2
59/15	3.93	1.13	13.3

There are only 10 useful gear ratios, and here they are:

Teeth pedal/wheel	*Ratio*	*Ratio of ratio*	*Change (%)*
35/27	1.30		
35/23	1.52	1.17	17.4
35/20	1.75	1.15	15.0
46/23	2.00	1.14	14.3
59/27	2.19	1.09	9.3
35/15	2.33	1.07	6.8
46/17	2.71	1.16	16.0
46/15	3.07	1.13	13.3
59/17	3.47	1.13	13.2
59/15	3.93	1.13	13.3

4. If h is the number of white hard hats and m is the number of managers, the relationship can be expressed as $\frac{5}{7}m = \frac{3}{4}h$. Therefore, on 'cross-multiplying', $5 \times 4m = 3 \times 7h$ so $20m = 21h$. So if $m = 21$ then $h = 20$. Therefore the ratio of white hard hats to managers is 21 : 20.

Averages of averages are not always the average

Averages do not always present the same message, as this scenario shows.

A company has three divisions, each of which are keen to ensure there is an appropriate balance between able-bodied and disabled employees. The table shows how many were recruited, compared with the number who applied:

Division	*Able-bodied*	*Disabled*
A	7 out of 10 = 70%	1 out of 1 = 100%
B	15 out of 20 = 75%	4 out of 5 = 80%
C	2 out of 10 = 20%	7 out of 20 = 35%

In each case, the recruitment rate, in percentage terms, was higher for the disabled group compared with the able-bodied group. If we take the average of these percentage recruitment rates, they are:

	Able-bodied	*Disabled*
Whole company	55%	72%

This reinforces our earlier view that the recruitment rate, in percentage terms, was higher for the disabled group. However, if we take the company as a whole, that is add the figures for each group, able-bodied and disabled, then quite a different picture emerges.

	Able-bodied	*Disabled*
Whole company	24 out of 40 = 60%	12 out of 26 = 46%

The disabled recruitment rate is now well below that of able-bodied employees. This is an example of Simpson's paradox, where the result of a whole is different from the result of its parts. This may come about when the whole is made up of separate parts which are uneven in size. For example, in Division A there are 10 able-bodied and one disabled applicant; however, the one disabled applicant causes an apparent recruitment rate of 100%.

So which is correct? The answer is that they all are, within the limitations of each data set. However, the one that is most meaningful is often the one that uses the most data. So in this case the company's recruitment policy is best reflected in the company's full set of data.

The Simpson's paradox can be seen graphically in this example. Two groups of students are taught together. Group A are quite bright and get good marks in spite of poor attendance. Group B are quite the opposite; they attend well but get poor marks. Here are the results from their final exam:

Group	*Exam marks*	*Attendance*
A	100	10
A	94	9
A	93	8
A	90	7
A	87	6
B	30	20
B	24	19
B	17	18
B	13	17
B	10	16

In the graph, Group A are in the cluster at the top left and Group B are the cluster at the bottom right. Looking at each

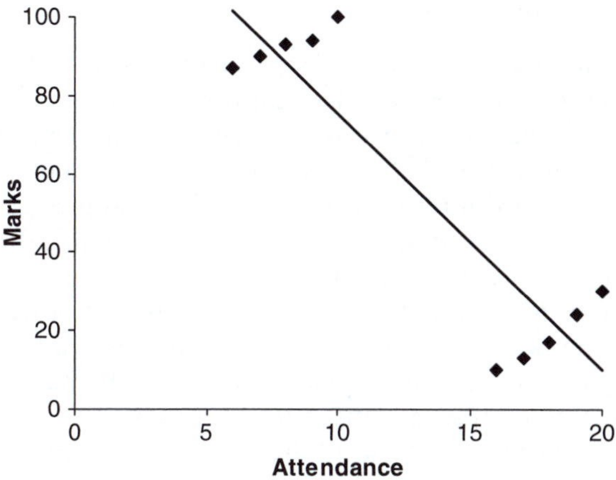

group separately, there is an obvious positive correlation between attendance and marks; better marks come with better attendance. The line joining the two groups is the graph trend line and is based upon all of the data. A coefficient of correlation shows whether there is a link between marks and attendance. For this data set it is –0.9. This suggests that the best marks are obtained with the worst attendance. This anomalous result comes from putting together two very disparate sets of data.

Something to think about

Can you explain how an engineer who changed jobs decreased the average IQ of both the company he left and the one he joined?

He was brighter than the average employee of the company he left but joined a company of more intelligent people than himself.

Summary of the main points

- Single digit numbers are easy to comprehend. On the other hand, very large and very small quantities are difficult to visualise. Expressing a quantity in different units, years instead of seconds or grams instead of tonnes, may help. Changing the number into standard form may make it easier to appreciate the relative magnitudes of disparate values, for example, the size of an atom to the size of a planet.
- Some quantities may be expressed very precisely and others less so. Quoting a value to a greater precision than can be justified implies that a statistic has greater worth than it actually has. If you quote a statistic to less precision than it really has, you throw away some of the value of that statistic.
- No measurements are perfect. Errors in measurements translate into errors in quantities derived from those measurements; a calculated volume may be the product of a height, a breadth and a width, or the rate of fuel consumption of a vehicle may be calculated using the fuel dispensed and distance travelled.
- The implications of interest rates are not always appreciated. Rates can be expressed in different ways and when the rates are high or applied over a long period, the amount to be paid back compared with the amount borrowed may be considerable.
- Average is a word that can have several different meanings, such as the mean, the mode or the median. Which is the most appropriate average to calculate will depend on the purpose for which it is to be calculated and on the data available to calculate it.
- Even the mean may have different meanings. When calculating the mean of a set of data which is most appropriate: arithmetic, weighted, geometric, harmonic or root mean squared? It all depends upon your application.
- A ratio compares the size of one quantity to the size of another or the size of a part to the size of the whole. Applying ratios successively is the same as applying the product of those ratios. Successive ratios cannot be added; to do so might even produce

impossible answers. If inflation reduces the value of your savings by 5% each year that does not mean that after 20 years you are left with nothing; you actually have about 36% of what you started with.

- Finding the average of averages of components of data can lead to surprising and counter-intuitive results, especially when compared with the average of the data set as a whole.

5 False assumptions and false logic

Powder of sympathy and commonplace coincidence

We often look for a connection between cause and effect. In mathematics that is often justified by finding a statistical correlation between two parameters. But we must be careful not to make unwarranted assumptions.

Before GPS and other electronic aids to navigation were invented, the navigator's big problem when using the sun or stars to find position at sea was that of precisely determining time. The dilemma is that time and longitude are related. If you know your longitude it is possible to determine the time using observations to the stars (as astronomers may do), and if you know the time you can use the stars to determine your longitude; which is a navigation problem. But if you know neither you are lost. Prior to the advent of reliable and precise clocks that could be taken to sea for long voyages, the main method of navigation was to sail north or south until the desired latitude was achieved, which is relatively easy to measure, then sail due east or west until, often literally, you hit land.

One proposed 'magical' way to carry time to sea was to take a wounded dog on the ship. On shore a bandage which had been applied to the dog's wound was dipped in 'powder of sympathy', ground up copper sulphate crystals, at exactly mid-day. At the same instant the dog at sea was supposed to yelp with pain, thereby telling the crew the time.

This seems quite ridiculous now but may have been taken seriously in the 17th century. It could happen that the dog yelped close to midday but that would be just coincidence. Coincidences do happen, but can be explained by the laws of probability. In the case of the yelping dog, you need to know how often the dog yelps and what is meant by 'close to midday'. If the dog yelps randomly 24 times a day and 'close to midday' means 5 minutes either side of midday then there is a $\frac{1}{6}$ chance that the dog will yelp close to midday. However, since the dog also yelps at other times as well, there is nothing that can be concluded about the time for each occurrence of the dog's yelp. So if by chance the dog yelps at midday that is pure coincidence.

Coincidence is when two events that are significant happen at the same time but without any obvious reason for their doing so. When coincidence happens in real life we generally think of it as something remarkable. An unlikely event has happened to most of us at some time or other. A school friend, unseen for 30 years, turned up in the house where I was staying in Nairobi in Kenya once. How unlikely was that? It was very unlikely that he should be in that house at the same time as I was, but much less unlikely that our paths should cross somewhere and at some time in this life. It is even less unlikely that I should unexpectedly meet an unspecified old school friend somewhere, someday.

Coincidence is less remarkable than you may suppose. What is the probability that you and I were born on the same day of the week? $\frac{1}{7}$, but that is hardly amazing. Similarly, the chance that we were born in the same month is $\frac{1}{12}$, again nothing very significant about that. The chance that we were born on the same day of the year is approximately only $\frac{1}{365}$.

But consider these coincidences the other way around. You probably know many people; neighbours, family, work colleagues or people at college or university, for example. Let's say 253 all together. What, then, is the probability that you were born on the same day as at least one of them? This is easier to compute as one minus the probability that none share your birthday. The probability that any particular person does not share your birthday is $\frac{364}{365}$. The probability that they all do not share your birthday is $\left(\frac{364}{365}\right)^{253} = 0.500$. So the probability that at least one of your acquaintances shares your birthday is $1 - 0.500 = 0.500$; perhaps less than you imagined.

How many people do you need to gather together to be sure there is a 50% chance that at least two have the same birthday? We can express this problem the other way round as finding the number of people where there is a 50% chance that no two have the same birthday. If there are n people, the chance that no two have the same birthday is $\frac{365!}{(365-n)! \times 365^n}$. If this is not greater than 0.5 then the maximum value for n is 23. Therefore, you need to gather 22 people other than yourself.

You cannot prove that there is a causal connection merely because there is a coincidence. Back to the beginning, if the dog yelps and it happens to be midday that does not prove that the cause of the yelping has something to do with the powder of sympathy. However, leaving ethical considerations aside, if you were to carry out a series of experiments with a number of wounded dogs and did find a significant correlation between yelps and the time at which the bandages were dipped in powder of sympathy then that would lend considerable support to the hypothesis that there was a causal connection. Since no research has ever produced such evidence we may presume there is no connection and so, unsurprisingly, there is no evidence that powder of sympathy works.

Something to think about

Is it remarkable that when you meet a stranger on a train, they know someone that you also know? What is the likelihood that you and I have at least one common acquaintance in the UK if we each know 1000 people randomly distributed among the 60 000 000 or so people that there are in the UK?

> For any person, taken at random from the UK, there is a $\frac{1000}{60\,000\,000} = 0.000\,0167$ probability that you know that person. Therefore, there is a 0.999 9833 chance that you do not know that person. The chance that you do not know any of my 1000 acquaintances is therefore $0.999\,9833^{1000} = 0.983$. That means that there is a $1 - 0.983 = 0.017$, or a 1.7%, chance that you do.

False assumptions about data

Some scientific investigations are carried out by researchers of integrity whose only interest is to determine the truth. Others may then seek to present the results for their own less honourable purposes. Consider this statement:

> Only 20% of people questioned objected to the building of a new bypass.

That statement may be entirely true of course, but there is nothing in it that gives confidence in its honesty. Some potential flaws are obvious but others may be more subtle.

It appears that since only 20% objected there must be 80% who are in favour. What defines 'objected'? It could be that 20% were in outright opposition and that those who 'don't know' or did not respond or had a neutral opinion are included by default with those who did not object. If the range of possible responses included 'object', 'strongly not in favour', 'mildly not in favour' as well as more positive responses, then that begs the question as to how 'strongly not in favour' and 'mildly not in favour' have been counted.

How big was the sample? Have 1000 people been canvassed, or just 100 or only 10? Generally, the larger the sample, the more reliable are the statistics. If it had been only 10, what is the probability that this result could have been achieved with a population that is equally split between those for the bypass and those against? The answer is approximately 0.055, which suggests that the conclusion from such a small sample is rather weak.

Who was asked? In other words, how representative are they of the whole population who might have been asked? Perhaps those nearer the proposed bypass might have a different view from those on the other side of the town, so was a wide sample used from across the town?

What question was actually asked? Was it:

> Are you in favour of solving the town's congestion problem with a new bypass? Or
>
> Do you wish to preserve our countryside rather than having it destroyed for the sake of a few town traders?

Both questions are biased. The first assumes there is no other solution to the problem. The second presents a negative view of the proposal and is therefore unbalanced. A 'yes' to both questions sends a very different message.

Who was undertaking the survey? Are they looking to make the survey give the answer that favours them? In other words, is there likely to be bias in the conduct of the survey?

Have the results been 'cherry-picked'? Are the results as presented biased because some of the data that was collected has not been used in forming the research conclusion?

Expectation based upon previous performance

We all have unfounded expectations from time to time and then are disappointed. Some things in life have near mathematical certainty but others do not.

How long will it take you to get to work or college? You may have made the journey many times and found that it takes, on average, 1 hour with a standard deviation of 5 minutes. The time taken depends on the level of traffic. Can you now safely make the assumption that this will be the case in the future? For the purpose of getting to work, the worst case of 1 hour and 10 minutes will probably do; only rarely are you late. But in life, very occasionally something disastrous happens. You have an accident, perhaps, and have to wait for the police to let you go. The hourly bus is full and does not stop for you. There is heavy traffic which grinds to a halt.

Whatever the purpose of your journey, the likelihood of delay will be the same. So if the purpose of the journey differs, how will you react? For example, you have an interview for a high profile and well-paid job, or you are the star witness in a major trial, or you have an important examination. Clearly, in each case, you need to be well on time. With a standard deviation of 5 minutes for your journey time that would suggest that if you budget for a standard journey of 1 hour then for 4.6% of journeys you will be 10 minutes late, 0.3% of journeys you will be 15 minutes late, 0.006% of journeys you will be 20 minutes late, etc. So since you cannot easily afford to be late, you decide to accept the risk of 0.006% and allow 1 hour and 20 minutes for your journey. Is that a wise choice?

You have to make a decision and accept some risk; being 100% certain is never possible no matter how early you start out. Each 5 minutes you leave early greatly reduces your chance of being late according to the statistics of the normal distribution. But since the calculations on which you have worked out your average time and its standard deviation have never included a 'disaster', how valid are they? The answer is that they are not and given the importance of your journey you would be well advised to ignore the statistics and get in with an hour or two to spare. Take a book or some work with you so as not to waste your time.

Do you play the horses or follow the dogs?

Imagine that your favourite horse has come 5th, 4th, 3rd and 2nd in its last four races. This is an arithmetic progression of places and so the next term is 1st. So do you feel confident to place your life savings on this horse? If you did you would be mistaking the successive numbers of an arithmetic progression for placing of the horse. Previous performance in something that has so many variables and uncertainties as a horse race cannot be reduced to a string of related digits. What if there was another horse in the race with the same profile; should that also come first? Even if your horse did come first, this logic would suggest that it will come 0th next time; whatever that means.

Some people like to play the stock market and some people even make money on it through 'wise' investments. To make money you need to identify which stocks are rising and which are falling. If you have the graph of a stock you can see how it has performed in the past. If there has been a consistent and steady rise, can that be taken as evidence of the stock's future trajectory? Sadly not, or we would all be rich. Like the horse in the previous example, stock prices do not follow some magical formula or trend but respond to a large number of factors, some of which relate to the company, some to the market in general, and some movements in price seem to have no logic at all. Almost all stock market related 'investments' caution that 'past performance is no guarantee of future results', and that is exactly what it means; it is a self-delusion to think otherwise.

Global warming is the new big issue, or is it? In the early days there was much speculation and little data to back one view or the other. While there seem to be some worrying trends with respect to weather patterns, there is still a range of predictions being offered for future climates. There are now masses of data, and there are some very sophisticated computer models, but there is no absolute certainty for outcomes. This is because although mathematical modelling is at the heart of predictions the relationships described by those models are empirical and therefore subject to error in the data on which they are based. But as time goes on it is likely that the models will improve, thereby making predictions more reliable. We know from practical experience that, with short-term weather forecasting, the shorter the term of the prediction the more accurate it tends to be.

Something to think about

What factors do you think make political election predictions unreliable?

Factors may include: changeability of the electorate's opinion prior to an election, events and factors which come to light between the announcement of an election and the election itself, non-random selection or bias in selection of the population sample, selection of people who will actually vote, and truthfulness of the population sample.

Fallacies of argument

Here is a paragraph about the work of the structural engineer.

> A factor of safety used in the design of a structure is the ratio of the load that causes failure to the maximum designed load that may be applied. Its application makes sure that a structure will not fail because of uncertainty in the design process. As good engineers we must all have faith in British scientific investigation to determine the physical properties of materials that we work with. We should follow the methods developed by the great scientists and engineers like Einstein, Rutherford, Newton, Galileo and Brunel. When designing structures, if we do not apply factors of safety of three or greater we are doomed to create structures that will fail and will be held accountable for the consequences. No one wants to be accused of risking the lives of young children. All the best structural engineers use minimum factors of safety of three; therefore that is the correct value to be used. A factor of safety of three is better than a factor of safety of two and any engineer who uses factors of safety of two is guilty of negligence. Before employing a structural engineer make sure that he confirms that he no longer uses factors of safety of two. Engineers traditionally use a factor of safety of three. A factor of safety of three is good but a factor of safety of two is bad. Make sure the designer has done everything possible to ensure safety. Studies have shown that people who use weak bridges will die.

> Everybody knows of the well-documented case of the young engineer who redesigned a bridge abutment with a reduced factor of safety, only to see the bridge collapse. If reduced factors of safety are permitted then we will all be at risk from failing structures. Do you agree with this, and if not, why not?

You may not agree with what has been presented but is the argument logical? Let's pick it apart.

As good engineers we must all have faith ... As good engineers the last thing we must rely on is *faith* in anything which is unproven.

... *in British scientific investigation* ... Why only British scientific investigation? This implies that we should not accept scientific investigation done in any other country but it does not present any argument to justify why only British scientific investigation is appropriate.

We should follow the methods developed by the great scientists and engineers like Einstein, Rutherford, Newton, Galileo and Brunel ... Methods developed by the great names of science may or may not be appropriate for addressing modern problems. All discoveries are subject to re-examination and re-evaluation; understanding is never set in stone.

When designing structures, if we do not apply factors of safety of three or greater we are doomed to create structures that will fail ... What is the evidence for this? This is a statement presented as a universal truth without justification.

... *will be held accountable for the consequences* ... This is an appeal to support the previous argument by threatening you if you do not agree.

... *accused of risking the lives of young children* ... This is an appeal based on the use of sentiment rather that logical argument.

All the best structural engineers use minimum factors of safety of three ... Do they? And for every design? This is a statement presented as truth when it is not so.

... *therefore that is the correct value to be used* ... The fact that one person uses something for their own purposes does not mean that it is correct for what you are doing.

A factor of safety of three is better than a factor of safety of two ... Better in what way? It implies that the structure will be safer. However, it will be much heavier and if it has to be built on poor ground it may then be subject to subsidence. This is a blanket statement presented as a universal truth when it is not.

... *any engineer who uses factors of safety of two or less is guilty of negligence* ... This is only true if the factor of safety is not appropriate for the particular design.

Before employing a structural engineer make sure that he confirms that he no longer uses factors of safety of two ... This is dubious advice for reasons already given, but this statement has implications that all structural engineers have been previously operating dangerously. It is an argument akin to the prejudiced 'have you stopped beating your wife?' type of question.

Engineers traditionally ... Tradition is, of itself, no argument to justify future action.

A factor of safety of three is good but a factor of safety of two is bad ... This is another universally fallacious truth; it all depends upon the application.

... *has done everything possible to ensure safety* ... This is impossible as there is always more that can be done; there comes a point where the expense of further action is unreasonable.

... *studies have shown* ... What studies? There is no citation to any reference.

... *that people who use weak bridges will die* ... We will all die sometime. The unsupported implication is that weak bridges are a cause of death to all who use them.

Everybody knows ... No they don't. This is an appeal based upon shaming the reader into accepting an argument based upon assumed superior knowledge of others.

... *well-documented case* ... No citation to any reference again.

... *of the young engineer who redesigned a bridge abutment with a reduced factor of safety, only to see the bridge collapse* ... The unsupported implication is that the reduced factor of safety was the cause of the collapse. It could have been anything.

If reduced factors of safety are permitted then we will all be at risk from failing structures ... An unqualified statement presented again as a universal truth.

Do you agree with this, and if not, why not? ... This is an attempt to coerce you to agree by bullying. Agreement requires no action but disagreement calls on you to justify yourself.

The fraction pi?

Pi, π, is the ratio of the circumference to the diameter of a circle. In school you were perhaps told that π was $\frac{22}{7}$. That was a convenient half-truth; convenient in its simplicity but a very limited truth in that it is an approximation that is good to only three significant figures. So is there perhaps a more complicated fraction that gives π exactly? The ancient Greeks erroneously thought so.

$\frac{22}{7}$ can be re-expressed as the first part of the simple continued fraction $3 + \frac{1}{7}$. The next improvement is $3 + \frac{1}{7+\frac{1}{15}} = \frac{333}{106}$ which is good to 10^{-4}. At least this has a few prime factors in both numerator and denominator. It may be rewritten as $\frac{3^2 \times 37}{2 \times 53}$, which may aid cancelling if this approximation for π is used with other integers or fractions.

Better approximations are:

$\frac{355}{113}$ which is good to 10^{-6}

$\frac{103\,993}{33\,102}$ which is good to 10^{-9}

$\frac{5\,419\,351}{1\,725\,033}$ which is good to 10^{-13}

There are many approximations for π. Some of the more concise are $\frac{66\sqrt{2}}{33\sqrt{29}-18}$ and $\frac{9801}{2206\sqrt{2}}$ which are good to eight significant figures and $\sqrt[4]{\frac{2143}{22}}$ which is good to nine significant figures.

Rather neater are these approximations:

$$\sqrt[3]{31}=3.1414 \quad 6^{\frac{23}{36}}=3.14161 \quad e^{\sqrt[3]{1.5}}=3.14154$$

$$\sqrt[5]{306}=3.14155 \quad \frac{7^7}{9^9}=3.14157 \quad \sqrt[11]{294204}=3.14159264$$

$$\sqrt[15]{28\,658\,146}=3.141\,592\,653\,8$$

In each case, the last significant figure shown on the right-hand side of the equals sign is in error:

A decimal value for π to 1000 decimal places is stated in chapter 10.

But the point of this section is that π is an irrational number, which means there is no fraction that can express it exactly. There is also no square or other root, or rational power of a rational number or logarithm, that can give an exact value for π.

Something to think about

e is the number defined such that when you differentiate e^x with respect to x and evaluate the derivative at any point, the value of the derivative is equal to the value of the original function. In perhaps more familiar terms $\frac{d(e^x)}{dx}=e^x$. The approximate value of e is 2.718 281 828 459 045 235 360 287. Like π it is also irrational. Can you find any smart approximations for it?

There are a number of approximations for e to be found on the internet. Here are a couple of the neater ones.

$$\sqrt[6]{\pi^4+\pi^5}=2.71828181 \quad \text{and} \quad 3-\sqrt{\frac{5}{63}}=2.7182819$$

But you might ask: are they really that neat? The first approximation is good to eight significant digits and its approximation involves a total of seven digits, symbols and letters: 4, 5, 6, π, π, + and $\sqrt{\ }$. The second approximation is good to seven significant digits and its approximation involves a total of six numbers and symbols: 3, 3, 5, 6, − and $\sqrt{\ }$. In short, it takes almost as much ink to write either $\sqrt[6]{\pi^4+\pi^5}$ or $3-\sqrt{\frac{5}{63}}$ as 2.7182818.

Zeno's arrow

Zeno, who came from Elea in Greece, lived in the fifth century B.C. Much of his work was summarised in a series of paradoxes which presented logical arguments that came to illogical or contradictory conclusions. One fundamental argument that he employed was that time and space are infinitely divisible and hence length and time could be broken down into an infinite number of parts.

The paradox of Zeno's arrow is one of his best known. It runs like this. Suppose that you and I stand some distance apart and I shoot an arrow at you. If my aim is good and you stay still, will you be struck by the arrow? Zeno would have us believe this is not so, and this is his argument.

> Let us imagine that the distance between us is 32 metres. It does not have to be this distance but 32 is a good number because it is divisible by 2 several times and that makes the arithmetic in the explanation somewhat simpler.
>
> After the arrow has been fired it heads towards its target; that is you. Before it gets to you it must first travel half the distance. We can treat the whole distance as the sum of two halves, each 16 metres long.
>
> By a similar argument, before the arrow can get from the halfway point to you it must first get to half of that distance;

> that is, a distance of 8 metres from the halfway point which is also 8 metres from you.
>
> But as there is now 8 metres from the arrow to you, before the arrow can get to you it must travel half of the remaining distance, 4 metres, to the point which is 4 metres from you.
>
> There is now 4 metres from the arrow to you, so before the arrow can get to you it must travel half of the remaining distance, 2 metres, to the point which is 2 metres from you.
>
> Now the distance is 2 metres from the arrow to you. Before the arrow can get to you it must travel half of the remaining distance, 2 metres, to the point which is 1 metre from you.
>
> ...
>
> And ... by now, hopefully you see where this argument is going. No matter how small the final section of the arrow's journey is, it can be split into two equally sized halves ... *ad infinitum*.

So the lengths of the sections in metres are 16, 8, 4, 2, 1, 0.5, 0.25 ... *ad infinitum*. If you have an infinite multiple of any positive number the product is also infinity, no matter how small the positive number may be. So since there is an infinite number of distances the sum of them all must also be infinity. We have proved that the distance is infinite?!

If you also look at this problem in terms of time where the arrow is travelling at a constant speed, let's say 32 metres per second, then since the distance between us is 32 metres that would imply that the time the arrow takes to get to the target will be 1 second. But before the arrow gets to you it must get half way, which will take half the time, $\frac{1}{2}$ second. The next section, being half as long, will take half of that time, $\frac{1}{4}$ second. The times for subsequent sections will be $\frac{1}{8}$, $\frac{1}{16}$, $\frac{1}{32}$, etc. of a second. Since the distance is broken down into an infinite number of sections then there is also an infinite number of periods of time. All of the times are positive and none are zero.

Again, we have an infinite multiple of positive numbers so the product is also infinity, by the same argument as we applied to the infinite number of distances. Since there is an infinite number of periods of time the sum of them all must be infinite. The arrow takes forever to fly so it can never reach you.

Likewise, if it takes an infinite amount of time for the arrow to reach you and the distance between us is finite, then, since speed is calculated as distance divided by time, $\frac{32}{\infty} = 0$, the arrow must be travelling at 0 metres per second. Since its speed is zero then it can never leave the bow and you are doubly quite safe.

So what went wrong with the logic?

The times for each section follow a geometric progression. That is, there is a constant ratio, r, of $\frac{1}{2}$ between each of the times; $\frac{1}{2} \times \frac{1}{2} = \frac{1}{4}$, $\frac{1}{2} \times \frac{1}{4} = \frac{1}{8}$, $\frac{1}{2} \times \frac{1}{8} = \frac{1}{16}$ and so on. The first term of the series, a, is $\frac{1}{2}$ and the number of terms in the series is infinite, ∞.

The formula for the sum of the first n terms in a geometric progression is $a\left(\frac{1-r^n}{1-r}\right)$.

In this case $n = \infty$, so the sum of the terms in this series is $\frac{1}{2}\left(\frac{1-\frac{1}{2}^{\infty}}{1-\frac{1}{2}}\right)$. Since any positive proper fraction multiplied by itself an infinite number of times is zero, this reduces to $\frac{1}{2}\left(\frac{1}{\frac{1}{2}}\right) = 1$ second and this now fits with our common-sense expectation that you will be shot.

Do not try this at home!

Something to think about

1. The lengths of the sections between us were 16, 8, 4, 2 metres, etc. Using the formula for the sum of the terms in a geometric progression, can you say what the total distance is?
2. In Zeno's paradox of Achilles and the tortoise, the athlete Achilles runs a race against a tortoise. Since Achilles knows he can run faster than the tortoise he gives the tortoise a head start. When Achilles gets to where the tortoise was when he, Achilles, started he finds that the tortoise has moved on. Achilles now runs to where the tortoise is but when he gets there the tortoise has moved on again. And again Achilles now runs to where the tortoise is but when he gets there the

tortoise has moved on. And ... so on. Therefore, Achilles can never catch up with the tortoise. Correct?

1. 32 metres
2. Obviously not. Use similar arguments to those of Zeno's arrow to show that Achilles will soon catch the tortoise and hence overtake it.

Summary of the main points

- Coincidences occur. However, the fact that one has occurred to you is not something remarkable in itself. It may be possible to estimate the likelihood of a given event happening *before* the event occurs. However, the likelihood of an event occurring after it has happened is always 1.
- Conclusions drawn from surveys are quite invalid unless the survey was undertaken with rigour and integrity. It is very easy, deliberately or inadvertently, to create bias in the results of a survey which in turn may lead to false conclusions.
- We often look for patterns in life and, having found them, make assumptions that they will repeat or not as the case may be. Random events are just that, random. From time to time patterns will occur in the outcomes of events but if the events are random then no conclusions can be drawn about future occurrences.
- False logic is often presented to justify doubtful arguments. Science and engineering must rely on proof, not faith. The origins of research, of themselves, do not indicate the quality of that research. Past methods and understanding are always subject to review and revision. Claims always need to be justified; there are very few universal truths. Condemning disagreement is not an argument in support of truth. An argument cannot be based on sentiment, tradition, bullying or coercion; there must be logic behind it. Statements presented

as truth, but without support, have little credibility. More or bigger is not always better without justification in context. Evidence in support of an argument will usually need an appropriate citation. Conclusions must always be supported by evidence.

- In the practical world some numbers must be approximated. e and π have specific values but are irrational; that is, they cannot be expressed exactly as a fraction. In decimal form these numbers have no end.
- The 'proof' of the paradox of Zeno's arrow, although intriguing, is false. Using the mathematics of the geometric progression, summation of the successive times or distances associated with the arrow's flight shows they are finite.

6 Lies, damned lies and statistics

Cause, effect and dubious correlation

Statistics is about drawing meaning from numbers. As the saying goes, 'there are lies, damned lies and statistics', meaning that the unscrupulous and the ignorant can make numbers mean what they want them to mean when it suits their purpose. Correlation and cause are often mistaken; indeed correlation is often erroneously used to explain cause. Correlation in statistics means the interdependence of two things, that is, the tendency of one factor to be associated with another. For example, medical costs tend to be correlated with age. The reason, the cause, is that as we get into old age our bodies start to give up and need more help to keep them going. Cause means that to some extent at least, outcomes are predictable. However, although cause may create correlation, correlation does not of itself explain cause.

There is a well-established link between smoking and lung cancer. Studies have certainly shown that the average life expectancy of someone who smokes is less than the life expectancy of someone who does not. However, to say that smoking *causes* lung cancer is a very dubious statement since, to prove this, it would be necessary to show that *all* people who smoke *will* get lung cancer. This is not true, although people who smoke are more likely to get it. A cause may lead to a correlation but the correlation of itself does not prove the cause.

All cigarette packs in the UK must, by law, have one of these two warnings; *smoking kills* or *smoking seriously harms you and others around you*. But are these statements both true? Medical research has established the correlation between smoking and early death in the first case and between smoking and harm in the second. However, if you smoke you will not necessarily die from it and you will not necessarily seriously harm yourself and others, although you are more likely to do so.

So which of these statements is true?

> Violent computer games cause violent behaviour.
> Poor housing causes crime.
> Sports cars cause more accidents.
> Gun ownership causes murder.

The answer is none of them. There seems to be a correlation between the use of violent computer games and violent behaviour, but for violent computer games to cause violent behaviour you would have to show that most users of violent computer games engage in violent behaviour. Likewise, poor housing does not cause crime but it would be argued that those who live in poor housing may be more likely to commit crimes. The poor housing is not the cause, but it does create the conditions whereby someone with a criminal tendency is more likely to offend. Sports cars do not cause more accidents, but drivers who travel fast are more likely to have an accident and sports cars are designed to travel fast. It may be that aggressive people are more inclined to buy fast cars. There are many careful and responsible gun owners but unfortunately there are also some who seek to own a gun with the express purpose of using it in criminal activity.

For those in the media here lies a problem; a correlation does not make for a dramatic headline but a cause does. Often the media will misrepresent correlation as cause. If the claim of a cause is qualified with *may* or *can* or some other statement reflecting a link that is probable rather than absolute, that would make for more responsible journalism. For example:

Violent behaviour may be induced by playing violent computer games.
High crime rates are associated with areas of poor housing.
The insurance premiums for sports car drivers are higher than those for other drivers because they tend to have more accidents.
Murderers have a higher rate of gun ownership than non-murderers.

These statements may be less exciting but are much easier to justify.

Something to think about

Which of these statements are clearly untrue?

1. Giving blood saves lives.
2. Wearing seatbelts saves lives.
3. Life expectancy in Cornwall is higher than the national average. Cornwall is the home of pasties and clotted cream. Pasties and clotted cream are good for you.

1. This statement is not true because lives are not saved by the mere act of giving blood. However, blood is a necessary commodity for operations which in turn do not have death as an outcome. Therefore, the selfless act of giving blood allows others to prevent untimely death. In that sense, giving blood is good but it takes more than that to save lives.
2. If you have a car accident and are wearing a seatbelt you are less likely to die. In that case wearing a seatbelt increases your chance of survival and therefore could be said to save lives. But of course wearing seatbelts does not save **all** lives.

3. This is an entirely spurious correlation presented as a cause. Current Office of National Statistics data seems to show that life expectancy in Cornwall is a little higher than the national average, based on local authority rankings. Cornwall is famous for its pasties and clotted cream. However, it does not seem at all likely that overindulgence in these cholesterol-laden foodstuffs will promote a long life. It is more likely to be something to do with the friendliness of the Cornish, the milder climate and lower crime rate that enables people to live a bit longer.

Loaded questions in surveys

Surveys are often used to find out about people's attitude to, or opinions of, ideas or proposals. Surveys may be by interview, or more often by questionnaire. Conclusions are then drawn from the data, using statistical methods to interpret them. The quality of the conclusions that you can draw will depend upon the quality of the data, which in turn depends upon the quality of the questions that are asked.

Questions should be simply structured and use easily understood words and phrases that all respondents can understand. Here are two questions asking pretty much the same thing. Which do you think will give the most reliable set of answers?

> Since starting to read this tome have you formed a generally favourable opinion of its ability to give you a pleasurable reading experience? Yes/No.
>
> Have you enjoyed this book so far? Yes/No.

One question should seek only one answer. If your question is really two then respondents cannot separate the two parts and therefore the answers may be meaningless. Questions with the word *and* are often ambiguous.

Do you think that buses and taxis should be given priority in built-up areas? Yes/No.

It is often better to break the question into two:

Do you think that buses should be given priority in built-up areas? Yes/No.
Do you think that taxis should be given priority in built-up areas? Yes/No.

A respondent may be confused if the meaning of the question is unclear, that is, the respondent cannot be sure what is actually being asked.

Is wood a better building material than steel? Yes/No.

This is so ambiguous as to be meaningless. Better in what way? It could mean is it stronger, more sustainable and more aesthetic or cheaper. Not all structures could be built in either material. A better and more specific question might be:

Do you think that wood used for structural members in the construction of single-storey residential buildings is a cheaper building material than steel? Yes/No.

Questions need to be answerable.

Do your lecturers know their subjects? Yes/No.

This question is almost impossible to answer unless you know almost as much as the lecturers concerned. The answers are also going to be variable because although your lecturers will know something, can you say whether they know enough in the context of what they are teaching you? It is also a very difficult question to answer if you have several lecturers with different capabilities. This question might be better if asked for each lecturer, asked in the context of your own learning and if a range of opinions rather than a simple yes/no answer is offered.

A leading question is one that directs the respondent to a particular answer. There is a presumed truth in its text that coerces. The next question presupposes that you have had an accident and that you should have reported it.

> The last time you had an accident why didn't you report it?

Similarly, there is only likely to be one answer to:

> Do you think that your employer is paying you too much? Yes/No.

A loaded question might be one for which there can be no right or favourable answer. For example, in the next questions either response implies a level of moral guilt or irresponsibility:

> Do you still drink before driving home? Yes/No.
> Have you started giving blood yet? Yes/No.

Hardly a research question, but there is the old classic:

> Have you stopped beating your wife?

Something to think about

Are these good or bad questions?

1. How often do you break the law when driving? Once a day/once a week/once a month.
2. Before completing this survey please give a few details about yourself.

Age ...	Income £ ...	Address ...
Religion ...	Race ...	

3. How do you get to work? Tick the box that applies to you.

walk ☐ cycle ☐ car ☐

1. A poor question. It is loaded in that any of the answers offered amounts to a confession. It does not offer the possibility of a 'never' answer. Respondents are unlikely to be truthful, which thereby invalidates your survey. A better way to ask this would be to invite the respondent to make a value judgement about others rather than themselves.

   Think of someone you know well but do not name him or her. To the best of your knowledge how often do they break the law when driving? At least once a day/once a week/once a month/never.

2. A dreadful opening to a questionnaire. This is very personal information and many respondents would baulk at completing the questionnaire. Ask yourself if you really need this kind of information. Even if you had it what would you do with it? To be meaningful in any statistical sense you would need to aggregate it in some way, so why not make it anonymous to start with? If you really need this information give ranges of answers such as:

   Age 20–35/36–50/51 or more
   Income £ 10K–20K/20K–40K/40K–80K/more than 80K
   First 2 letters of your postcode ...

   As for the religion and race questions—give a standard list of alternatives.

3. Another poor question. This presupposes that all respondents go to work; how should any who do not go to work respond? It does not give any alternatives for those who do go to work and use any form of public transport.

Making 'statistics prove' by biasing the sample

The phrase 'statistics prove' is a contradiction in terms. The subject of statistics is about getting meaning from numbers; statistics is the mathematics of uncertainty and variability. Uncertainty can never prove anything in any absolute sense. However, statistics can tell us how likely something is to be true and in that sense statistics can allow us to support or reject a hypothesis.

When data has been collected for an investigation or research project, normally it should all be used unless there is a good reason associated with known deficiencies in the quality of the data. To do otherwise is to select arbitrarily a subset of the full data set. If this is done randomly or by accident then it may not seriously compromise the integrity of the analysis. However, much more serious is the deliberate rejection of data that does not support an argument that you wish to maintain. This is the use of statistics to create lies. If the hypothesis is only just supported to an appropriate level of significance by the data, then even minor tinkering can change an outcome.

For example, in a new manufacturing process where you plan to make special bolts it is important to make them as close to 10 mm in length as possible. Any bolt above 10.20 mm has to be rejected. It is in your interest to be able to report to senior management that the percentage of bolts that is rejected is low. The quality control consists of selecting bolts at random and measuring them. You do so, and here are the results, all in millimetres:

10.08	10.09	9.97	9.93	9.97	10.04	10.04	10.07	10.17	10.12
9.87	9.99	10.08	10.03	9.96	10.11	9.97	10.00	10.16	9.91

You calculate, correctly, the mean as 10.028 and the standard deviation of the sample as 0.0826 mm. Therefore, bolts which are $\frac{10.20-10.028}{0.0826} = 2.08$ standard deviations above the mean will be rejected. With reference to the normal distribution that is 1.86% of all bolts will be rejected. You know that this will be unacceptable to your boss so you consider how to improve the data. You know of course that it is false to change any data but on looking at it you note that although none of the bolts is greater than 10.20 mm you might be able to enhance the overall result by not using a few of the measurements. Here is what you decide to use:

10.08	10.09	9.97	9.93	9.97	10.04	10.04	10.07
9.87	9.99	10.08	10.03	9.96	9.97	10.00	9.91

The mean is now 10.000, which is exactly what you would like, and the standard deviation of the selected sample is 0.0654 mm. The bolts which are $\frac{10.20-10.000}{0.0654} = 3.06$ standard deviations above the mean will now be rejected. This represents only 0.11% of any batch of bolts and is a far more encouraging statistic to report. With some small selectivity of the data you have produced a much better result; or have you?

On the basis of the unselected data it appears that about 19 bolts per thousand will fail, but your revised results suggest it will be only 1 per thousand. What will happen in practice? Because the original data is representative of all bolts, since the bolts to be measured were selected at random, it is the original data that most truly reflects the 'population' of bolts. Massaging the data by being selective only produces a false result and this deception will be exposed as soon as the production line starts to run.

Something to think about

You wish to work out how many hand-made tiles are required to make a frieze on a long wall. You measure a few tiles at random to get an estimate. Your measurements in millimetres are:

83.6 86.7 84.5 85.1 89.9

This set of measurements has a mean of 85.96 mm and a standard deviation of 2.48 mm. Later you make another random set of measurements, they are:

87.6 80.1 83.9 83.4 84.5 79.7 81.9 87.1 82.8 81.0

This has a mean of 83.20 mm and a standard deviation of 2.69 mm.

Although the standard deviations of both sets are similar, the means are a bit different. Which set of measurements should you accept as definitive? How can you get the best possible estimate of the size of a tile?

The first is a smaller set but has a smaller standard deviation, which suggests a closer agreement within the set. The second set is twice the size of the first set but has a slightly larger standard deviation, which suggests a less close agreement within the set. So which is definitive? The answer is neither, because a random selection from the whole population of tiles is just that, a random selection. We can only estimate the mean of the set from the mean of a sample unless we measure every tile. To get the best possible estimate of the size of the average tile use all the data that is available to you. So putting both sets together you will find that the mean is 84.12 mm and the standard deviation is 2.87 mm.

Ignoring factors that influence an outcome

In the last section, we saw that if you are selective with the data you have already obtained then you deliberately bias your results; it is the research equivalent of cheating. However, you can still affect the outcome of your investigation by not taking account of all the relevant factors, which can happen inadvertently.

Here is a simple example: say you are investigating the social costs of smoking by analysing existing health data. Therefore, you look at costs to the National Health Service of smoking-related diseases across different age ranges, genders and economic groups, and compare results for smokers and non-smokers. You also weigh this against the tax revenues from tobacco. From this you could decide whether smoking is of net financial benefit to the nation, or otherwise. Or can you? Such an investigation would be simplistic since it does not take account of all the financial effects of smoking.

In your investigation you would almost certainly find that smokers tend to die younger than non-smokers. One consequence of this is that their burden on the NHS would be less protracted and that would reduce the cost to the nation.

We all see dangerous driving by other people; but never do it ourselves, of course. In deciding what to do about unsafe driving, a highways authority might choose to commission a survey to find people's perceptions of what are the most antisocial activities that need to be addressed. A questionnaire could be sent to households at random, asking questions designed to rank in order the types of driving behaviour in others which causes the greatest fear. The returns might then be correlated with the gender, age, income and vehicle ownership of the respondent. Assuming that there were sufficient returns, how complete would such a survey be? That rather depends upon whether all the relevant factors have been taken into consideration in the survey, which in turn is very difficult to assess.

The statistical analysis would be likely to show if an irrelevant factor has been included because that factor would turn out not to be statistically significant. On the other hand, identifying a factor which then turns out to be significant is rather more problematic. In this example, how significant would geographical location be? Might it be that city dwellers view speeding or jumping red lights as more or less significant than rural inhabitants? If there were big differences in response, that might suggest that variations in regional policing are more appropriate than a blanket policy for the whole country. Ignoring the effects of geography might lead to quite the wrong approach to finding the best solution to the problem of dangerous driving.

Many lecturers consider that motivation among students is in decline. The 'evidence' often cited for this is the apparent decline in levels of attendance at lectures. An investigation might be undertaken to see whether this is so, recording attendance in successive years at the same lectures across a number of courses at the same institution. Would such an investigation be valid? Have all the appropriate factors been taken into account?

It may well be shown that levels of attendance are reducing, but is that truly evidence for declining motivation? There may be other factors that affect attendance at lectures. Modern universities and colleges are making greater use of other means of delivery, especially electronic means such as virtual learning environments, so that more students are now learning by means other than lectures. So, if other learning methods are a factor, ignoring them will lead to the wrong conclusions.

Wood is a construction material that has been around since man moved out of the caves. But wood is not a homogeneous material like steel. To be able to design buildings with wooden structural components it is necessary to know how wood behaves. One of the important properties of a construction material is its crushing strength. In devising an experiment to determine the crushing strength of wood what factors should be taken into consideration? Obviously, the size and shape of the samples will be important. Temperature and moisture content may also be significant. But are these the only factors? Ignoring the orientation of the wood is likely to affect the result; crushing strength along the grain may be different from crushing strength across the grain. The occurrence of knots or decay may also have a significant effect. But probably the most important missing factor is the type of wood itself; oak and pine are not likely to have the same properties.

Something to think about

You are tasked with investigating the design of tents for a variety of recreational, domestic and industrial purposes. What do you think might be the engineering factors that need to be considered in evaluating a tented structure? You might start by grouping your factors under headings such as: the purpose of the tent, the properties of the membrane, the support structure (how it stays up) and the anchorage (what holds it in place).

A comprehensive list of the engineering factors that should be considered for the design and use of tents are considered at http://handle.dtic.mil/100.2/ADA137382. In summary, this site suggests, among others:

Purpose: size, style, location, heating, ventilation, entrance and exit.

Membrane: tensile strength, fabric tear, wear, reinforcement, degradation, transmission of light, resistance to airborne chemicals, fire resistance, waterproofing, weight of fabric and cost.

Support: frame design, frame shape, mechanical properties of the frame, failure mechanisms, strength, interface between frame and membrane, ease of assembly.

Anchorage: guy ropes if used, connection of anchorage to membrane or support, ground anchors, adjustment mechanism.

Stock market predictions—do they work?

What makes the stock market go up and down and can we predict these events? Some would say we can. There are, generally, two groups of analysts: fundamental analysts and technical analysts. Fundamental analysts are those who look at the fundamental worth of a company in comparison with its stock market value and attempt to forecast the future direction of the company on the basis of its record of profits, its cash flow, its assets and the return on those assets. If the stock market valuation underestimates the company's worth investors would tend to buy, and if overestimated would tend to sell. Technical analysts would largely disregard company information and focus on what the share price has done in the past and what it is doing now, with a view to predicting what it will do in the future.

However, there are other factors which lie outside any company, such as those factors that affect the market as a whole. If investor confidence in general falls, then so will the stock market. Investor sentiment may be affected by political events which in turn may be driven by military action. False confidence, prompted by over-optimism, may temporarily make the market rise and corrections may then make it fall. Sometimes those falls are large, as in the global banking crisis of 2008, for example. Major external factors, which even the fundamental analyst looking inwards on a company may not see, could have a significant effect on a company's ability to trade.

Technical analysts will use various statistical tools to analyse past performance in an attempt to predict future movement. There is some element of truth in the assertion that patterns of behaviour in stock prices repeat themselves. It is one thing to identify two similar patterns but quite another to predict when a repeating pattern will occur. There is an awful lot of noise surrounding weak signals.

In spite of the difficulties, there are many predictions around. One fundamental question to ask is why anyone would make a prediction that others could benefit from. If you have discovered an expectation that a particular stock will rise, why not keep quiet and take advantage yourself? No doubt many do and perhaps they make a profit. One advantage of broadcasting a prediction of a rise in value is that, if it is believed, then people will go out and buy the stock. If there are more people buying than selling, the stock's price will rise and the prediction will come true. This is known as a self-fulfilling prophecy.

You probably get many spam emails. There is a group of them that try to persuade you to invest in particular stocks—often with a tag line that the stock is undervalued and the price will rise dramatically in the next few days. This will be true if you and others believe it. However, once you have bought and the price has risen the originator of the spam email sells his stock and the price comes crashing down again, probably leaving you with a loss! Similarly, if you are subscribing to a forecasting service, it may not be as unbiased as you would hope. After all, why would someone sell you good advice when they could make more money by acting on it alone?

So whose advice should you follow? Advisers who have a consistent long-term track record may have something to offer but the advice will probably be expensive. Those who have made infrequent but spectacular gains should be treated with caution, because they will almost certainly have made significant losses as well. Just as estate agents have the reputation for talking up the housing market, many financial advisers are probably similarly motivated.

Something to think about

If it was that easy to predict stock market movements then everybody would be doing it and making huge profits. But for every winner there is also a loser; everybody cannot be both a winner and a loser. So how can you be sure of making a profit on the stock market?

> The straight answer is that you cannot. All stock market-based products are obliged to display the warning that 'past performance is not indicative of future results'. The only way you could guarantee making a profit would be to have information about the company concerned and to use it. That would be called 'insider trading' and is strictly illegal.

Going to the dogs

Some people enjoy a night at the dogs or a day at the races. Others take their gambling more seriously and look on their betting as an 'investment'. A few may even make a regular profit; it can only be a few otherwise the bookmakers would be out of business very quickly. It would be a big mistake to think that this, as a means of income, is suitable for the average punter. The only way you can regularly expect to beat the bookies is if you can spot opportunities where the probability of a particular dog or horse winning is better than that suggested by the odds being offered. This requires detailed knowledge of the participants in the race.

Let's be clear about the terms involved. Probability is likelihood. A probability of an event occurring of 0.1 means that given an infinite number of random recurrences of the event, 0.1 or 10% of the time the event occurs. So for an infinite number of independent events that have the probability of 0.1 for a positive outcome, 10% of the events will actually have a positive outcome.

Odds, on the other hand, are the ratio of the winnings (excluding the stake) to the stake, offered by the bookie. So, for example, the odds based solely on the probability of the event that the roll of an unbiased die will show a 6, will be 5 to 1.

Odds are related to probability, p, by:

$$\text{odds} = \frac{p}{1-p}$$

Since the probability that the die will show a 6 is $= \frac{1}{6}$ then the odds that the die will show a 6 is:

$$\text{odds} = \frac{\frac{1}{6}}{1-\frac{1}{6}} = \frac{1}{6-1} = \frac{1}{5} \text{ which is the ratio of } 1{:}\frac{1}{5}$$

which is 5:1.

Strictly, these are the 'odds against'; there are 5 unfavourable outcomes and 1 favourable outcome. Sometimes odds are expressed as 'odds on' which is the ratio of favourable outcomes to unfavourable outcomes; 1:5 in this example.

Expectation is the average return for your unit stake and depends on the true probability of the event and the odds offered by the bookmaker. For example, if the probability that the spotted dog will win is $\frac{1}{5}$ but the bookmaker offers odds of 3:1 then the expected return will be $\frac{1}{5} \times \frac{3+1}{1} = 0.8$ so you can, on average, expect to have 80p back for each £1 you bet. In other words, on average, you will lose 20p every time you make such a bet. Since the bookmaker sets the odds to ensure a profit on every race the punters, overall, make a loss.

Here is an example based upon the odds offered by one particular bookmaker for a particular horse race. The question now is how much would you need to bet to ensure a return of £100? For each horse you would need to place £100 divided by the 'odds plus stake'.

Horse	*Odds*	*Odds + stake*	*Stake*
Tucker	9:4	3.25	£30.77
Nanton	7:2	4.50	£22.22
Run to Space	9:2	5.50	£18.18
Yachvili	11:2	6.50	£15.38
Knight Valliant	8:1	9.00	£11.11
Montchara	10:1	11.00	£9.09
Super Cris	20:1	21.00	£4.76
Hunting Haze	18:1	19.00	£5.26
Devil Water	33:1	34.00	£2.94
Rory's Star	33:1	34.00	£2.94
Modarab	80:1	81.00	£1.23
		TOTAL	£123.90

So for each £123.90 you bet you are guaranteed £100 back whichever horse wins. Therefore, your expectation is $\frac{100}{123.90} = 0.807$.

The only sensible bet is one where the expectation is greater than 1.00. So just about every other commercially offered betting opportunity is not a wise investment because you must expect to get back less than you 'invest'.

But suppose you need to make some money and have no choice but to gamble. Which strategy is most likely to achieve your objective given of course that nothing can guarantee success? So let's say that you need to get £50 and have only £10 to bet on the dogs. Suppose also that the bookie is offering odds of 4 to 1 for all dogs so that for each £1 bet on a winner you end up with £5.

Should you put everything on a single dog or should you make lots of smaller bets and so spread the risk?

Strategy 1. Let's say that you bet your £10 on one dog. If the dog wins up you get £50 and your objective is achieved; any other dog and you have nothing and walk away a loser. There are six dogs altogether of which only one is the winner, therefore your chance of winning by this method is $\frac{1}{6} = 0.167$. Not good; but at least the uncertainty is short lived; you have the outcome after just one race.

Strategy 2. Start by betting £5 on your favourite dog. If you win you get £25 to make a total of £30. If you lose you still have £5. From here on you bet the lesser of half your current asset or the minimum amount necessary to achieve a total asset of £50. For example, if you currently have £30 then you need a further £20. So bet £5 because, at 4 to 1 odds, if you win you will get your stake of £5 plus your winnings of £20 and so have a total of £50. If you get to your target of £50, then stop betting. If your assets become less than £1, give up and go home.

So does this improve your chances of walking away with £50? To find out we need to investigate the probabilities of all possible outcomes, and this is quite complex. Alternatively, the 'method' can be tested by running the strategy many times and comparing the number of good outcomes, where you get your £50, with bad outcomes where you walk away with less than £1. In one such trial of 100 000 runs of strategy 2, the number of wins was 12 601. This equates to an overall chance of winning of 0.126.

Strategy 2 is clearly worse than strategy 1. The chance of achieving the objective of having £50 is 1 in 6 with strategy 1 and about 1 in 8 with strategy 2.

In fact, in any game of chance where the odds are in favour of the bookie, as they almost always are, a strategy that involves betting with earlier winnings is, in the long run, less likely to be successful than one that involves placing one single large bet. Why should this be so? With the odds in favour of the bookie, he or she makes, on average, a profit on each bet. So with a greater value of all the bets, on average the greater will be the profit for the bookie and therefore the greater the loss for you. In other words, there is a smaller chance that you will achieve your objective.

Something to think about

Can you think of any ways of making an honest living from betting on the dogs or horses?

Here are two, rather obvious ideas.

Study the form of your animals so that you become significantly more knowledgeable than the other punters and the bookmakers.

Become a bookmaker.

But don't think that any betting *strategy* will work.

How will your luck change?

Luck and probability are often confused. Probability may be stated as the numerical level of certainty that an event will happen, in which case the probability of related events may also be calculated. Luck, however, relates to a belief in the chance of good or bad fortune when there is no rational reason to believe it will occur. Luck can only favour someone who will benefit from it. Some believe that good or bad fortune may be influenced by the time and place of birth, in which case life must to some extent be predestined. Others may see luck as being influenced by quasi-random factors outside their own control such as which way you turn at a junction. Arbitrary decisions are assumed to have life-changing consequences.

Whereas luck is just the effect of random and capricious uncertainty, some may see good fortune as an outcome of actions. For example, number 7 is considered by some to be a lucky number. One day I took a number 7 bus and on my seat I found a £50 note, therefore my good fortune was caused by taking a bus with a lucky number. Therefore, I always take a number 7 bus because I will soon find more money on the bus. This is the false logic of attributing consecutive events with causality: two events occurred, therefore we assume they were related; I took a number 7 bus therefore I found the money so if I take number 7 buses in the future I will find more money, it is assumed.

There are some, with more or less credulity, who seek to influence outcomes by superstitious action. Good fortune, they assume, will follow finding a four-leaf clover, possession of a rabbit's foot or carrying an artefact which is assumed to have benign influence. Bad luck may follow walking under a ladder, breaking a mirror or stepping on the cracks in the pavement. Although in these cases there may be something rational in that, respectively, something may be dropped on your head, you may be cut by the broken glass or you might twist your ankle on the uneven pavement.

So if these are irrational ways of supposedly influencing your good or bad fortune how can you use mathematics to enhance your fortune in everyday life? Some gamblers have systems that they believe in.

If I always use the same numbers in the national lottery I must win one day. There is a limited truth in this belief since any number is as good as any other and it makes no difference which numbers you choose. The limited truth may be verified by considering what the chance of consecutive no-wins is. To keep the numbers reasonably simple let's suppose that you have to choose a four-figure number from 0000 to 9999, that is 10 000 possibilities. Your lucky number is 1234 which you use every time you play. The stake is £1 and the single prize is £10 000; so, overall in this case, the lottery organiser is not expected to make a profit. The chance of winning is $\frac{1}{10\,000}$ and therefore the chance of not winning is $\frac{9999}{10\,000}$. The chance of two consecutive no-wins is $\left(\frac{9999}{10\,000}\right)^2$ and the chance of n consecutive no-wins is $\left(\frac{9999}{10\,000}\right)^n$. It is not until $n = 6932$ that the chance of n consecutive no-wins is less than 0.5. Indeed the chance of no-wins in 10 000 consecutive plays is 0.368, so even if you happened to win on the next play you would still have lost more than you staked altogether. So how many plays must you make to be certain of a win? The answer is an infinite number of plays; but no matter how rich you are, your wealth is finite. To reduce your risk of not winning to 1 in 100 you would still need over 46 000 plays.

Another gambler's ploy to try to avoid loss is to double the stake until there is a win; the double or quit scenario. The idea is that for an initial stake there is either a win or, if a loss, then that is recovered by successive stakes. If a gamble of £1 is on the basis of the toss of an unbiased coin the chance of a win is $\frac{1}{2}$ and the chance of a loss is $\frac{1}{2}$. The chance of no loss after the second toss, total now staked £2 + £1, is $\frac{1}{4}$. The chance of no loss after n tosses, total staked £$(2^n - 1)$, is $\frac{1}{2^n}$. So how many plays will it take to guarantee no loss at all? In mathematical terms that is when $\frac{1}{2^n} = 0$ which is when $n = \infty$. It will take an infinite number of plays with an infinite stake. So to try to save losing £1 you must be prepared to risk more than the universe.

So back to luck; if following heavy gambling losses you say that chance forbids that this will go on forever and your luck will turn, you are only fooling yourself. In successive unrelated events past performance is no guide to future events if the events are random. A run of 50 successive heads has no bearing on the next toss of the coin; it is just as likely to be heads as tails.

The only way to bias the chance of winning in your favour is if you can spot an opportunity where the likelihood of winning is greater than the odds offered. This will not happen in a casino or with any knowledgeable organisation that runs games of chance. If it did, they would soon be out of business. Gambling opportunities where there is subjective setting of the odds such as horse racing may occasionally provide opportunities for the very knowledgeable punter. Card counting may help to give you an edge, but casinos watch for gamblers who do so.

Something to think about

In the UK's national lottery, to win the jackpot, you must select the six winning numbers from the 49 numbers available. The chance of picking the right combination is 6 out of 49 for the first number, 5 out of 48 for the second number, 4 out of 47 for the third number and so on. Therefore, the chance of picking all 6 numbers is:

$$\frac{6 \times 5 \times 4 \times 3 \times 2 \times 1}{49 \times 48 \times 47 \times 46 \times 45 \times 44}$$

$$= \frac{1}{13\,983\,816} \quad \text{(or 1 in almost 14 million).}$$

There are 13 983 816 possible combinations of six numbers picked randomly from 49. So:

1. If you are determined to have at least a 0.5 chance of winning the jackpot in one draw how many different selections must you make for that draw?
2. If you make one selection a week for how long must you plan to make your weekly selection to have at least a 0.5 chance of ever winning the jackpot?

1. To have a 0.5 chance of winning the jackpot you need 0.5 of the numbers; 6 991 908.
2. The chance of not winning in any 1 week is $\frac{13\,983\,815}{13\,983\,816}$. The chance of n consecutive no-wins is $\left(\frac{13\,983\,815}{13\,983\,816}\right)^n$. If this is 0.5 then:

$$\left(\frac{13\,983\,815}{13\,983\,816}\right)^n = 0.5$$

$$\text{So } n = \frac{\ln 0.5}{\ln 13\,983\,815 - \ln 13\,983\,816} = 9\,692\,842$$

With one entry a week it will take you 186 400 years to get only a 0.5 chance of ever winning. Good luck!

Expectation and high/low probability of low/high outcome

And finally, in this section, one last thought of the risky business of gambling. This is about what you *can* expect to win. In this context, 'expectation' means the sum of the probabilities of all possible outcomes multiplied by the gain associated with each of the outcomes.

$$\sum p(x) \times x_o$$

where $p(x)$ is the probability of event x and x_o is the outcome of that event.

So, for example, in a game of dice you are required to bet £4 and your winnings are £1 times the number rolled. All numbers are equally likely so the probability of rolling each of the six numbers is $\frac{1}{6}$. The outcome for the roll of a 1 is £1, for the roll of a 2 is £2 and so on. The probability of loss of the £4 stake is 1. This means that the expectation is:

$$\frac{1}{6} \times £1 + \frac{1}{6} \times £2 + \frac{1}{6} \times £3 + \frac{1}{6} \times £4 + \frac{1}{6} \times £5 + \frac{1}{6} \times £6 - 1 \times £4$$
$$= -£0.5$$

An expectation is not the most likely outcome and may not even be a possible outcome, as in this case. However, it is the average return if the event is run a large number of times.

So let's consider the expectation associated with different events. In the section ***Going to the dogs*** the odds offered on a horse race were shown. Now let's say we had assessed the probabilities that each of the horses would win, before the race of course. The bet placed on each horse is £1. Notice that the sum of all the probabilities must be exactly 1 if it is certain that there will be a winner and that there will be only one winner.

So with a £1 bet on all the horses the expectation is a loss of £3.32. In fact, assuming we assessed those horses' probabilities correctly there is only one horse worth backing and that is Nanton. Overall, the expectation per £1 'invested' is –£0.33.

Horse	*Winnings (including bet)*	*Prior assessed probability of a win*	*Product of winnings and probability*	*Expectation*
Tucker	£3.25	0.24	£0.78	–£0.22
Nanton	£4.50	0.30	£1.35	+£0.35
Run to Space	£5.50	0.14	£0.77	–£0.23
Yachvili	£6.50	0.10	£0.65	–£0.35
Knight Valliant	£9.00	0.07	£0.63	–£0.37
Montchara	£11.00	0.06	£0.66	–£0.34
Super Cris	£21.00	0.03	£0.63	–£0.37
Hunting Haze	£19.00	0.02	£0.38	–£0.62
Devil Water	£34.00	0.02	£0.68	–£0.32
Rory's star	£34.00	0.01	£0.34	–£0.66
Modarab	£81.00	0.01	£0.81	–£0.19
	TOTALS	1.00	£7.68	–£3.32

In a particularly simple lottery the only prize is £10 000 and 14 930 £1 tickets are sold. You buy one ticket so your expectation is:

$$\frac{1}{14\,930} \times £10\,000 - 1 \times £1 = -£0.33$$

Both events, the horse race and the lottery, show very similar expectations but the nature of the events is quite different. In the lottery the favourable outcome, a £10 000 win, is much greater than any possible outcome in the horse race. However, the chance of achieving it is very small. So given that the expectations for both events are very similar, which is the better event to bet on? The answer depends on what you are trying to achieve. If it is to have the best chance of a win, then go for the horse race. If it is to win a fortune use the lottery. But either way bet with the assumption that you will lose, but be surprised (and pleased) if you win.

Something to think about

What is the only event that the wise gambler will bet on?

The wise gambler will only bet if the loss can be afforded and then only on an event where the expectation is greater than 0. But can you find one? Probably not!

Summary of the main points

- Cause and correlation are two quite separate concepts. Cause may result in correlation but correlation does not, of necessity, imply cause.
- Surveys are used to find out opinions and facts. The output of a survey may be very dependent on the wording of the questions that are asked. Questions which are ambiguous or all-encompassing are unlikely to lead to meaningful conclusions. Leading and loaded questions will elicit biased answers.
- Being selective of research data might be used (by others!) to steer results towards a previously favoured conclusion. This is bad practice and conclusions won in this way will not be valid.
- Similarly, in trying to identify cause from correlation you may miss a significant cause if you have not identified and investigated all relevant factors, which would otherwise lead to a correlation.
- The price of stocks and shares can be volatile and unpredictable. There are some in the investment industry who have a reasonable understanding of how the stock market works. There are probably many others who would like you to think they have the required talent to manage your money. Ultimately, there is no certainty in prices and you should approach these kinds of 'investment opportunities' very cautiously.
- The betting industry is just that, an industry designed to show a profit from handling your wagers. So although you may have some wins, even spectacular wins, in the long run you will lose. If that was not the case the bookmakers would be going out of business.

- Luck and probability are quite different concepts. It may be possible to calculate the probability of a given future event, but whether that event will occur or not no one can calculate with certainty, unless the probability is 1 or 0 of course. Luck, however is an irrational belief in the likelihood of good or bad fortune. Luck has no place in mathematics or statistics.
- The expectation of an outcome and the likelihood of a particular outcome are quite different concepts. The expectation is simply the sum of the probabilities of all possible outcomes multiplied by the gain associated with each of the outcomes. The likelihood of an outcome is just the probability that the desired outcome will actually happen.

Part III

Some mistakes that we make

7 Errors in arithmetic

Miscarrying

Try this piece of simple mental arithmetic. Make sure you do it as mental arithmetic, don't use a calculator and don't sneak a look at the answer.

Start with 1000 and add 40
Next add another 1000
Then add 30
Add another 1000
Now add 20
Next add another 1000
Finally add 10
... and your answer is?

Did you come up with an answer of five thousand; Most people do. (I have written the number in words so that your eye is not drawn to it while following the instructions above.) If you did come up with five thousand then you got it wrong. Really! Yes. Go back and check. And if you still have five thousand, try it with a calculator and you should now get four thousand one hundred, the correct answer.

So why did you get it wrong (assuming you did)? Notice how the numbers are laid out in the instructions. By using different words in each line the numbers do not line up in the sense that units are not above units, tens above tens, etc. Also, the addition of thousands is interspersed with the addition of tens. When it comes to addition of the final 10 to the running total of the tens $(40+30+20)$, that is, 90, you were tricked into seeing the result with an extra 0 and hence making one thousand instead of one hundred. This was then added to the four thousand you have for the sum of the thousands so far, to make the erroneous five thousand.

Knowing your numbers

How well do you know your number bonds and 'times tables'?

At one time these were drilled into every child at an early age in school. Not knowing the relationship between numbers makes it as difficult to do sums as not knowing the sounds that letters make hampers reading. But calculators make us lazy. When presented with a simple problem how quickly do you reach for your calculator? There are a good many people who will not bother to attempt even the simplest sums without one. What they fail to realise is that it is often easier and quicker to do simple sums in your head but that requires you to know basic number relationships.

Try this quick quiz about number bonds without a calculator. All you have to do is identify which of the following sums are incorrect. If you time yourself, you can compare your speed with your accuracy.

Quiz 1

1. $3+2=5$	2. $3+8=11$	3. $7-11=4$
4. $27-18=19$	5. $9+7=15$	6. $13-7=6$
7. $19+17=36$	8. $37+28=55$	9. $24-17=7$
10. $25+27=42$	11. $28-9=19$	12. $73-37=36$

Here is another quick quiz, this time about simple products, in other words times tables. Once again, identify which of the following products are incorrect. Time yourself again, to compare your speed with your accuracy.

Quiz 2

1. $3 \times 4 = 12$	2. $4 \times 7 = 28$	3. $7 \times 5 = 25$
4. $6 \times 4 = 24$	5. $8 \times 12 = 76$	6. $6 \times 7 = 42$
7. $7 \times 12 = 94$	8. $8 \times 6 = 46$	9. $9 \times 7 = 64$
10. $3 \times 9 = 29$	11. $9 \times 5 = 45$	12. $7 \times 9 = 36$

If you had difficulty with this quiz you should review and improve your knowledge of the times tables. To be fluent with the use of numbers you need to know that nine eights are seventy-two, not just to be able to work it out but know this without having to think about it. Learning the times tables is rather like being an actor who has to learn lines for a television show. You need to know the tables so well that the answer comes out almost before the question has sunk in. So, for example, seven eights are ...?

Number bonds and times tables are the essential basics but there are other number relationships which are useful to be able to recall instantly. Powers of small numbers and factorials are handy. In the next quiz evaluate each of the expressions.

Quiz 3

1. 3^2	2. 2^3	3. $3!$	4. 4^2	5. 3^3	6. 5^2
7. $5!$	8. 11^2	9. 5^3	10. 2^5	11. 2^{10}	12. 3^5

The inverse process is just as useful. In the next quiz, identify the powers represented by these numbers. For example, 81 would be 3^4. Several have more than one answer. For example, 256 is 2^8, 4^4 and 16^2.

Quiz 4

1. 4	2. 8	3. 9	4. 16	5. 25	6. 27
7. 32	8. 36	9. 49	10. 64	11. 81	12. 100
13. 121	14. 125	15. 128	16. 144	17. 169	18. 196
19. 216	20. 225	21. 243	22. 256	23. 289	24. 324
25. 343	26. 361	27. 400	28. 441	29. 484	30. 512
31. 529	32. 576	33. 625	34. 676	35. 729	36. 784
37. 841	38. 900	39. 961	40. 1000		

Quiz 1 Questions 3, 4, 5, 8, 10 are incorrect.

Quiz 2 Questions 3, 5, 7, 8, 9, 10, 12 are incorrect.

Quiz 3

1. 9	2. 8	3. 6	4. 16	5. 27
6. 25	7. 120	8. 121	9. 125	10. 32
11. 1024	12. 243			

Quiz 4

1. 2^2	2. 2^3	3. 3^2	4. 2^4 & 4^2
5. 5^2	6. 3^3	7. 2^5	8. 6^2
9. 7^2	10. 2^6, 4^3 & 8^2	11. 3^4 & 9^2	12. 10^2
13. 11^2	14. 5^3	15. 2^7	16. 12^2
17. 13^2	18. 14^2	19. 6^3	20. 15^2
21. 3^5	22. 2^8, 4^4 & 16^2	23. 17^2	24. 18^2
25. 7^3	26. 19^2	27. 20^2	28. 21^2
29. 22^2	30. 2^9 & 8^3	31. 23^2	32. 24^2
33. 5^4 & 25^2	34. 26^2	35. 3^6, 9^3 & 27^2	36. 28^2
37. 29^2	38. 30^2	39. 31^2	40. 10^3

Appreciating scale

Which is the bigger scale 1:1000 or 1:5000?

The temptation is to say that it is 1:5000 because it involves the larger number, but that would be to misunderstand what is meant by scale. A plan with a scale of 1:1 gives a representation of the ground at life size. This permits very fine detail to be drawn but it is unlikely that the plan would be of much use as it would be very unwieldy. Such a plan of the room you are in would be as big as the room.

Mapping scales are constructed so that a relatively large area of ground can be represented on a manageable piece of paper. So at a scale of 1:1000, a piece of land or a building is represented at $\frac{1}{1000}$ of its true size. This fraction could also be written in decimal format as 0.001. Notice the equivalence of the expressions 1:1000, $\frac{1}{1000}$ and 0.001. A ratio may be thought of as just an alternative way of writing a fraction or its decimal equivalent.

So, if 1:1000 is the same as 0.001 then 1:5000 is the same as 0.0002 and since 0.001 is greater than 0.0002 then the scale 1:1000 is greater than the scale 1:5000.

At a scale of 1:1000 a building that is 100 m long would be 0.1 m long on the map. However, the same building would be 0.02 m at a scale of 1:5000. If you are drawing a plan or map then there will be several factors you need to consider when selecting the appropriate scale. Decide what the smallest object that you need to show is and how you are going to show it. If you wanted to show a rectangular manhole cover with sides of 0.55 m, and a square of side 5 mm (0.005 m) is the smallest that you can sensibly draw, then your smallest scale would be $\frac{0.005}{0.55} = \frac{1}{110}$ which is 1:110. Since this is a rather difficult scale to work with you would probably choose a scale of 1:100 as this will be more convenient and the scale is slightly larger.

Before metrication, scales were often chosen so that standard units on the ground would be represented by different standard units on the map or plan. For example, a scale of 1 inch to the mile would be 1:63 360 because there are 63 360 inches in a mile. Each inch on the map represents a mile on the ground. Nowadays, the nearest standard scale for mapping is 1:50,000; a slightly larger scale.

Quiz

1. Three different maps at scales of 1:2500, 1:12 500 and 1:7500 each show the same river. On which map does the river appear widest at the mapping scale?
2. In area, how many times smaller or larger will a building appear on a map of scale 1:1000 compared with a map of scale 1:5000?
3. A straight road is 3.2 km long. How long does this road appear on a map that is at a scale of 1:50 000?

4. Two hilltops are 234 mm apart on a map with a scale of 1:200 000. What is the true distance between the hilltops?
5. The distance between two buoys 25 nautical miles apart is shown on a nautical chart to be 926 mm. What is the scale of the chart? (1 nautical mile is 1852 metres.)

1. 1:2500
2. 25 times larger
3. 64 mm (0.064 m)
4. 46.8 km
5. 1:50 000

Negative signs

Here is a sum: $(-2)(-3)-4(-5)=-14$. True or not?

Negative signs seem to present problems to many students. The sum $5-3=2$ is straightforward because it may be interpreted as:

Positive 5, subtract positive 3, equals plus 2.

It could also have been interpreted as:

Positive 5, add negative 3, equals positive 2.

This last would have to be written as $5+(-3)=2$. The brackets are necessary as without them there would be two operators, + then −, adjacent to each other and that is not permitted in arithmetic or algebra. Actually, the sign − can have two meanings; either it indicates the negativeness of the quantity that follows it or it is an operator signifying that the quantity that follows is to be subtracted from that which goes before.

In the case of $+(-3)$ the signs are saying, add the quantity of negative 3. The effect of this is the same as subtracting the 3.

So $+(-3) = -3$. There are three other similar relationships and together they can be summarised as:

$$+(+3) = +3$$
$$+(-3) = -3$$
$$-(+3) = -3$$
$$-(-3) = +3$$

In the case of multiplication, the product of 2 and 3, which is 6, could be written as $2 \times 3 = 6$ or more elaborately as $+2 \times (+3) = +6$ to indicate the positiveness of all the quantities. However, if one of the terms in the product is negative then the result is negative. If both terms in the product are negative then the answer is negative negative which is positive. So the four possible products are:

$$+2 \times (+3) = +6$$
$$+2 \times (-3) = -6$$
$$-2 \times (+3) = -6$$
$$-2 \times (-3) = +6$$

Notice that the patterns of + and – are the same in both sets of four relationships. This leads to the oft-heard chant of:

Plus a plus is a plus
Plus a minus is a minus
Minus a plus is a minus
Minus a minus is a plus

This relationship also applies to quotients since $\frac{+2}{+3} = +\frac{2}{3}$, $\frac{+2}{-3} = -\frac{2}{3}$, $\frac{-2}{+3} = -\frac{2}{3}$ and $\frac{-2}{-3} = +\frac{2}{3}$.

Actually, if you have a string of terms to multiply and/or divide all you need to do is count the number of minus signs involved. If that number is odd the answer is negative. If the number is even the answer is positive. Likewise, a negative number raised to an even power will be positive and a negative number raised to an odd power will be negative.

But to return to the original problem; what is $(-2)(-3)-4(-5)$?

$$(-2)(-3)-4(-5)=6+20=26$$

Quiz

Evaluate the following expressions.

1. $2-1$	2. $1-2$	3. $3-(-3)$	4. $-3-3$
5. $-1-(-1)(-1)$	6. $(-3)(-5)$	7. $(-3)^2$	8. $(-2)^3$
9. $(-1)^{28}$	10. $(-1)^{937}$	11. $\dfrac{-16}{-4}$	12. $\dfrac{-6\times 8}{3\times(-4)}$
13. $-3\times\dfrac{-18}{-6}$	14. $\dfrac{-6^3}{(-2)^2}$		

1. 1	2. −1	3. 6	4. −6
5. −2	6. 15	7. 9	8. −8
9. 1	10. −1	11. 4	12. 4
13. −9	14. −54		

Confusing numerator with denominator

Here is a simple mistake: $(-5)\div(-3)=0.6$

But how long did it take to spot it? It is one of those mistakes where you try to justify what you see even though it is wrong. Following on from the last section, you confirm that a minus divided by a minus is a plus. You may also have seen it is a division sum and there are two numbers involved in the division; the 3 and the 5. Natural expectation may lead you to link the 3, 5, 0.6 and ÷. However, 0.6 is not the answer to the sum as presented; it is of course approximately 1.67.

This is a fairly trivial example, but the problem often appears when trying to divide one fraction by another, especially when algebraic quantities or dimensions are involved. For example, divide $\frac{2}{3}$ by $\frac{7}{9}$. This can be interpreted as $\frac{2/3}{7/9}$. If you are using a calculator then either you would work directly with fractions or you would need to enter these fractions as $(2\div3)\div(7\div9)$. Notice that without the brackets your calculator would interpret $2\div3\div7\div9$ as $\frac{2}{3\times7\times9}$.

Dividing one fraction by another can be confusing. One way to simplify the problem would be to convert the division to a multiplication. Instead of dividing by a fraction, multiply by its reciprocal, because the effect of these is the same. For example, dividing by 3 gives the same result as multiplying by $1/3$. Therefore, our problem of finding $\frac{2}{3} \div \frac{7}{9}$ may be rewritten as $\frac{2}{3} \times \frac{9}{7}$ which, with cancelling, leads to $\frac{6}{7}$.

Likewise, in algebra, $\frac{2a^2}{3b} \div \frac{a}{6b^3}$ may be rewritten as $\frac{2a^2}{3b} \times \frac{6b^3}{a} = 4ab^2$.

Quiz

Can you evaluate the following expressions and give your answers as fractions?

1. Divide $\dfrac{4}{7}$ by $\dfrac{11}{14}$
2. Divide $\dfrac{10}{21}$ by $\dfrac{15}{28}$
3. How many times does $\dfrac{8}{7}$ divide into $\dfrac{11}{14}$?
4. Divide $\dfrac{15}{91}$ by $\dfrac{33}{35}$

1. $\dfrac{8}{11}$ 2. $\dfrac{8}{9}$ 3. $\dfrac{11}{16}$ 4. $\dfrac{25}{143}$

Powers and roots

1. What is the mistake with this equation? $64^{-\frac{3}{2}} = \frac{1}{\sqrt[3]{64^2}} = \frac{1}{16}$

The essence of this problem is to understand what is meant by the index, $-\frac{3}{2}$, where 64 was raised to the power $-\frac{3}{2}$. Think of this power as the product of the minus sign, 3 and $\frac{1}{2}$. The minus sign indicates that the answer is the reciprocal of the same quantity raised to a positive power. Thus, $64^{-\frac{3}{2}} = \frac{1}{64^{\frac{3}{2}}}$.

The term $64^{\frac{3}{2}}$ can be expressed as $(64^3)^{\frac{1}{2}}$ or as $\left(64^{\frac{1}{2}}\right)^3$. Both are the same because of a law of indices that states that a base raised to a power, further raised to another power, is the same base raised to the product of the two powers.

Since $3 \times \frac{1}{2} = \frac{1}{2} \times 3 = \frac{3}{2}$, then $64^{\frac{3}{2}}$ can be expressed either way. But one way may be easier to compute, though of course they will both give the same answer. Remember that a power of $\frac{1}{2}$ is the same as a square root, so:

$$\left(64^3\right)^{\frac{1}{2}} = \sqrt{64^3} = \sqrt{262144} = 512 \text{ and}$$

$$\left(64^{\frac{1}{2}}\right)^3 = \left(\sqrt{64}\right)^3 = 8^3 = 512.$$

It should be clear which was the easier route to follow, especially without a calculator.

So, back to the original problem; the correct answer should be:

$$64^{-\frac{3}{2}} = \frac{1}{64^{\frac{3}{2}}} = \frac{1}{512}$$

2. What do you understand by the term a^{b^c}? Is it a power of a power, as in the first part of this section, or is it a base raised to a term which is itself raised to a power, and is there a difference? The problem, and hence the ambiguity, is most easily explained using a numerical example: 2^{3^4} Treated as a power of a power it is evaluated as:

$$\left(2^3\right)^4 = 2^{12} = 4096$$

But treated as a base raised to a term which is itself raised to a power it is:

$$2^{(3^4)} = 2^{81} \approx 2.4 \times 10^{24}$$

The two results are vastly different; illustrating the fact that $\left(a^b\right)^c \neq a^{(b^c)}$ and that therefore the term a^{b^c} is ambiguous.

It follows that $\left(a^b\right)^{-c} = \left(a^{-b}\right)^c$ but that $a^{(-b^c)} \neq a^{(b^{-c})}$ unless $a = 1$. If you are not convinced, try this out with some simple integers.

Calculators are likely to give 4096 as the answer to 2^3^4 because they treat the calculation as they find it, that is 2^3 = 8 then 8^4 = 4096. So in a calculation it is important that you make clear your meaning by the appropriate use of brackets.

Quiz

What are?

1. $64^{-\frac{2}{3}}$ 2. $-(16)^{-\frac{3}{4}}$ 3. $(-16)^{-\frac{3}{4}}$ 4. $-(27)^{-\frac{4}{3}}$ 5. $(-27)^{-\frac{4}{3}}$

Evaluate both possible solutions for each of the following expressions.

6. 4^{3^2} 7. 1^{1^1} 8. 2^{2^2} 9. 3^{3^3} 10. 4^{4^4} (this might be a challenge for some calculators)

11. Using the results from above, find real values for n that satisfy the equation $n^{(n^n)} = (n^n)^n$.

1. $\frac{1}{16}$

2. $-\frac{1}{8}$

3. There is no real solution because it requires the 4th root of a negative number.

4. $-\frac{1}{81}$

5. $\frac{1}{81}$; the solution is real because the cube root of a negative number is also a negative number. Your calculator may have shown 'Maths Error' because it treated $\frac{4}{3}$ as a decimal number having converted it to 1.333 333 333. It then tried to find a decimal power of a negative number. However, if you break the problem down into the following stages you get the correct answer.

$$(-27)^{-\frac{4}{3}} = \frac{1}{\left(\sqrt[3]{-27}\right)^4} = \frac{1}{(-3)^4} = \frac{1}{81}$$

6. 4096 and 262 144

7. 1 and 1

8. 16 and 16

9. 19 683 and 7.626×10^{12}

10. 4.295×10^9 and 1.341×10^{154}

11. $n = 1$ and $n = 2$ work, as the answers to questions 8 and 9; but what about $n = -1$ or $n = -2$?

    For $n = -1$ that gives $n^{(n^n)} = -1^{(-1^{-1})} = -1^{(-1)} = -1$ and $(n^n)^n = \left(-1^{-1}\right)^{-1} = (-1)^{-1} = -1$. So $n = -1$ is a solution.

    For $n = -2$ that gives $n^{(n^n)} = (-2)^{\left((-2)^{-2}\right)} = (-2)^{\left(\frac{1}{4}\right)}$ which is not a real number, but $(n^n)^n = \left((-2)^{-2}\right)^{-2} = \left(\frac{1}{4}\right)^{-2} = 16$; therefore $n = -2$ is not a solution.

Summary of the main points

- Mental arithmetic needs to be done carefully. Look for checks that you can make. For example, add a column of figures from the top then add the same column from the bottom to ensure you come to the same answer.
- An essential requirement for performing mental arithmetic successfully is to know your number relationships, such as number bonds, times tables and powers of small integers.
- As the saying goes, 'size is not everything', but in practical applications of mathematics you will need to understand representations of size. Scale is just the ratio of the size of an object's representation to the object's size in the real world.
- Working with negative values can be problematic. An odd number of negative values multiplied or divided together leads to a negative outcome and an even number of negative values leads to a positive outcome. 'Minus a minus makes a plus'.
- In a fraction, the numerator is on the top and the denominator is on the bottom. When the fraction is written as a division sum make sure you divide it the right way round.
- The notations for roots and powers are quite specific. A negative power has the same value as the reciprocal of the same base to the same positive power. For example, $3^{-2} = \frac{1}{3^2}$. A unit fractional power of a base is the base to the root of the denominator of the fraction. For example, $4^{\frac{1}{3}} = \sqrt[3]{4}$. A vulgar fractional power of a base is the base to the root of the denominator of the fraction all to the power of the numerator. It is also the base to the power of the numerator all to the root of the denominator. For example, $2^{\frac{3}{5}} = \left(\sqrt[5]{2}\right)^3 = \sqrt[5]{2^3}$.
- Beware of powers of powers. Notation like 6^{5^4} is ambiguous. It could mean $6^{(5^4)}$ or $\left(6^5\right)^4$; they are quite different.

8 Errors in algebra

Multiplying coefficients

Here is a mistake. $y \sin x + 4y = 4y \sin x$. Can you say what it is and suggest why it was made?

Algebra is a daunting subject for many students. Somehow, where you have the real-world experience of working with numbers you have a certain clarity. Working with variables is much harder because the meaning of what you are doing can easily become obscured. A variable is represented by a letter that stands for a quantity whose value is usually unknown, at least to start with. So if the equation $3b = 6$ represents the fact that three books cost £6, then the variable b is the cost of one book in pounds. In this case, the value of b would be found by transposing the equation to make b the subject. The mechanism for doing so in this case is to divide both sides of the equation $3b = 6$ by 3.

Often part of the process of transposition is to simplify an expression which is part of an equation. This in turn may be made up of a number of terms. So in the equation $v^2 = u^2 + 2as$, which comes from mechanics, the expression $2as$ has three terms, 2, a and s, one of which is a number, 2, and the others are variables, a and s.

If similar expressions can be added then there will be fewer of them. If expressions have terms in common, then a collection of expressions with common terms may be replaced by a product of factors. This in turn may lead to the possibility of cancelling in a fractional expression or dividing throughout to simplify an equation.

So, back to the plot ... what went wrong with $y \sin x + 4y = 4y \sin x$?

Here the student has tried to gather terms together but confused addition with factorisation. If you want to add expressions they must be the same except for any numerical coefficient. So, for example, you can add $2a$ to $3a$ because other than the numerical coefficients 2 and 3, respectively the rest, just a in this example, is the same. Adding the terms gives:

$$2a + 3a = 5a$$

If a is an apple this equation would say that two apples added to three apples makes five apples. Thus, the expression $2a + 3a$ may be replaced by the expression $5a$. It is true that this is a very simple example, but it is all about understanding the meaning of algebra.

On the other hand, if we wished to add $2ab$ to $3ac$ then the operation of addition is not open to us because the expressions $2ab$ and $3ac$ have more than just the numerical coefficients which are different; ab is not the same as ac. Another way to look at this is to identify the fact that in $2ab + 3ac$ there is a common term, a, in the two expressions $2ab$ and $3ac$. Thus, because the order of multiplication within an expression is unimportant, $2ab + 3ac$ can be rewritten as $a2b + a3c$ and further rewritten as $a(2b + 3c)$ because if you 'multiply out the brackets' you are multiplying each of the expressions in the brackets by the coefficient a. Think of brackets as a box; so $a(2b + 3c)$ would mean that you have a boxes and each box contains the same thing, $2b + 3c$. So if you tip out all a boxes you will have a lots of $2b + 3c$, which is $2ab + 3ac$.

Again, back to the plot ... The sum of expressions $y \sin x + 4y$ cannot be formed by simple addition because the expressions, $y \sin x$ and $4y$, have more differences than just their numerical coefficients. But they do have a common factor, y. Therefore, the sum of the expressions could be rewritten like this:

$$y \sin x + 4y = y \sin x + y4 = y(\sin x + 4)$$

which is now a product of the factors y and $\sin x + 4$.

Quiz

Can you simplify the following expressions?

1. $xy + 3yx$
2. $pqr + qrs$
3. $abc + bcd + cde$
4. $a + bc + def$
5. $\sin x + 2\cos y + 3\tan z$
6. $x\sin y + y\sin x$
7. $x^2y + y^2x$
8. $4\cos y + 5x\cos y$
9. $\sqrt{p} + \frac{1}{\sqrt{p}}$
10. $(p - q)(q + p)(p^2 + q^2)$

1. $4xy$
2. $qr(p + s)$
3. $c(ab + bd + de)$
4. No simplification possible.
5. No simplification possible.
6. No simplification possible.
7. $xy(x + y)$
8. $\cos y(4 + 5x)$
9. $\sqrt{p}(1 + p^{-1})$, but is this actually simpler?
10. $p^4 - q^4$

Incorrect operations

Under the pressure of examination it is so tempting to act before thinking. Even the simplest of sums can sometimes throw us. Look at this:

$$99y = 3 \text{ so } y = 33$$

How long did it take you to spot the error? Possibly it was longer than you would care to admit, and if so, why the momentary number blindness?

It is all about seeing what you expect to see rather than what you actually see. In the equation and its solution there is a product of 99 and y and there are three numbers, 3, 33 and 99. You will probably have immediately noticed an obvious relationship between those numbers; that $3 \times 33 = 99$ and hence been ready to accept 33 as the value for y. Of course, by now you will have worked out the true value of y, perhaps as follows:

$$99y = 3$$

$$\text{Leads to} \quad \frac{99y}{99} = \frac{3}{99}$$

$$\text{Hence} \quad y = \frac{1}{33}$$

So this was all about inadvertently multiplying when it should have been dividing.

$$\text{If } 7 - 3x = 4x \text{ then } x = 7.$$

If you agreed with that, it was probably because you saw the numerical coefficients of x as -3 and 4 and, since $-3 + 4 = 1$, linked the $1x$ with the 7 to make $x = 7$.

Not right, of course, because the correct answer is $x = 1$, as the following resolution of the problem shows:

$$7 - 3x = 4x$$

Adding $3x$ to both sides of the equation

leads to	$7 - 3x + 3x = 4x + 3x$
hence	$7 = 7x$
so:	$x = 1$

This time it was all about inadvertently subtracting when it should have been adding.

Quiz

Can you identify and correct the errors where they occur?

1. $\dfrac{x+3}{5} = \dfrac{2x+5}{3}$, therefore $x = \dfrac{16}{7}$
2. $7(y+5) = 2(3y+19)$, therefore $y = 3$
3. $\dfrac{3(x+2)(2x+3)}{6x^2+23x+8} = 1$, therefore $x = 5$
4. $\dfrac{a+b-c+d-e}{e-d+c-b+2a} = -1$. There is no unique solution because this is one equation in five variables.
5. $\dfrac{3}{5}(10x+15y+10z) = 3(2x+3y-3) + 3z$. There are no unique solutions for x and y but $z = 1$.

1. $x = -\frac{16}{7}$
2. Correct.
3. Correct.
4. Start by multiplying both sides of the equation by $e - d + c - b + 2a$ and then subtract $a + b - c + d - e$ from both sides of the equation. You are left with $0 = -a$. There is no unique solution for b, c, d and e but $a = 0$.
5. There are no unique solutions for x and y but $z = -3$.

Sum of bases raised to powers

Is this correct?

$$x^y + x^z = x^{y+z}$$

No; the rules of algebra with respect to indices have been confused. The writer has misunderstood the rule that if you multiply two numbers which have the same base, but each is raised to a power, then the result is the same base raised to the sum of the powers. In the equation above x is indeed a base and is the same base for all terms throughout. On the left-hand side of the equals sign, y and z are powers. However, the terms are added, not multiplied, so the multiplication rule does not apply. Consequently, the left-hand side, $x^y + x^z$, cannot be simplified.

We can verify that the equation above is wrong with this numerical example. Take the values $x = 2$, $y = 3$ and $z = 4$. Then

$$\begin{aligned} x^y + x^z &= 2^3 + 2^4 & \qquad x^{y+z} &= 2^{(3+4)} \\ &= 8 + 16 & &= 2^7 \\ &= 24 & &= 128 \end{aligned}$$

So, since $24 \neq 128$, then $x^y + x^z \neq x^{y+z}$.

Depending upon what you might do with the expression $x^y + x^z$, it may be useful to re-express it in the form of a product rather than a sum. In that case:

$$\begin{aligned} x^y + x^z &= x^y + x^{y+z-y} \\ &= x^y + x^y x^{z-y} \\ &= x^y\left(1 + x^{z-y}\right) \end{aligned}$$

For example, if $x = 2$, $y = 10$ and $z = 12$, then using $x^y + x^z = x^y\left(1 + x^{z-y}\right)$

$$2^{10} + 2^{12} = 2^{10}\left(1 + 2^{12-10}\right)$$

$$= 2^{10}\left(1 + 2^2\right)$$

$$= 1024 \times 5$$

$$= 5120$$

Quiz

Which three of the following are true?

1. $x^y x^z + x^z x^y = 2x^{y+z}$
2. $x^y y^z z^x = (xyz)^{x+y+z}$
3. $(xyz)^{x+y+z} = x^x x^y x^z y^x y^y y^z z^x z^y z^z$
4. $x^{(y+x)} y^{(y+x)} = (x+y)^{(y+x)}$
5. $x^{(y-z)} x^{(z-y)} = x$
6. $x^{(y+z)} y^{(x+z)} z^{(x+y)} = (yz)^x (xz)^y (xy)^z$

1. 1, 3 and 6 are true.
2. No simplification is possible.
4. $x^{(y+x)} y^{(y+x)} = (xy)^{(x+y)}$
5. $x^{(y-z)} x^{(z-y)} = x^0 = 1$

The logarithm function

Here is a mistake that is often made.

$$\log(2x) = 2\log(x)$$

It represents a misunderstanding of the property of logarithms. The logarithm of a number is the power to which its base has to be raised to be the same as the original number. For example, if the chosen base is 10 and the number for which we wish to find the logarithm is 1000, then to get the number 1000 we would need to raise the base to the power of three. That is $1000 = 10^3$, therefore the logarithm (to the base of 10) of 1000 is 3. Some consequences of this definition are that:

$$\log_a(bc) = \log_a(b) + \log_a(c)$$

$$\log_a(b^c) = c\log_a(b)$$

In both cases, the subscript, a, is the base of the logarithm, and b and c are numbers. If no base is shown, as in $\log(2x)$ above, then the implied base is 10.

If the logarithm of a number, x, to a base, a, needs to be expressed to another base, d, then from the definition of logarithms it follows that:

$$\log_d(x) = \frac{\log_a(x)}{\log_a(d)} = \log_d(a) \times \log_a(x)$$

and hence:

$$\log_d(a) = \frac{1}{\log_a(d)}.$$

So, back to the beginning of this section, and using the logarithm relationships above, the two halves of the mistaken equation could separately be written as:

$$\log(2x) = \log(2) + \log(x)$$

$$\text{but} \quad 2\log(x) = \log(x) + \log(x) = \log(x \times x) = \log(x^2)$$

Quiz

Which of the following relationships are true?

1. $\log(36) = \log(3) + \log(12) = 2\log(6) = \log(144) - \log(4)$
2. $\log(abc) = \log(a) + \log(b) + \log(c)$
3. $\log(ab)^c = c\log(a) + \log(b)$

4. $\log_2(3) = \log_3(2)$
5. $\log(a^b b^a) = b\log(a) + a\log(b)$

1, 2 and 5 are true.

3. Not true; $\log(ab)^c = c\log(ab) = c(\log(a) + \log(b))$
4. Not true; $\log_2(3) = \dfrac{1}{\log_3(2)}$

Square root of a sum

Can the square root of a sum be expressed like this?

$$\sqrt{x+y} = \sqrt{x} + \sqrt{y}$$

Here is a case of misunderstanding the square root function. A simple numerical example will show this cannot be true. If $x = 9$ and $y = 16$ then:

$$\sqrt{x+y} = \sqrt{9+16} = \sqrt{25} = 5$$

$$\text{but } \sqrt{x} + \sqrt{y} = \sqrt{9} + \sqrt{16} = 3 + 4 = 7$$

We can also see the inconsistency by squaring both sides of the original equation:

$$\left(\sqrt{x+y}\right)^2 = x + y$$

$$\text{but} \quad (\sqrt{x} + \sqrt{y})^2 = (\sqrt{x})^2 + 2\sqrt{x}\sqrt{y} + (\sqrt{y})^2 = x + 2\sqrt{xy} + y$$

Similarly:

$$\sqrt{x-y} \neq \sqrt{x} - \sqrt{y}$$

This may be verified with the numerical example where $x = 169$ and $y = 144$ because:

$$\sqrt{x-y} = \sqrt{169-144} = \sqrt{25} = 5$$

$$\text{but} \quad \sqrt{x} - \sqrt{y} = \sqrt{169} - \sqrt{144} = 13 - 12 = 1$$

Quiz

1. Are there any values for x and y which are both greater than zero for which the equation $\sqrt{x-y} = \sqrt{x} - \sqrt{y}$ is true?
2. Does $\sqrt{x+y}\sqrt{x-y} = \sqrt{x^2 - y^2}$?
3. Does $(\sqrt{x} + \sqrt{y})(\sqrt{x} - \sqrt{y}) = x - y$?
4. If $\sqrt{x}$, $\sqrt{y}$ and $\sqrt{x-y}$ are all positive integers greater than zero, what are the smallest possible values for x and y?

1. Yes. x takes any value and $y = x$. This leads to $\sqrt{x-x} = \sqrt{x} - \sqrt{x}$ or $\sqrt{0} = 0$.
2. Yes.
3. Yes.
4. $x = 25$ and $y = 9$.

Assuming a universal truth

The relationship that:

$$\sqrt{x \times y} = \sqrt{x} \times \sqrt{y}$$

is well known. You could illustrate it by using some suitable numbers, for example, $x = 9$ and $y = 25$.

$$\sqrt{x \times y} = \sqrt{9 \times 25} = \sqrt{225} = 15$$

$$\text{and} \quad \sqrt{x} \times \sqrt{y} = \sqrt{9} \times \sqrt{25} = 3 \times 5 = 15$$

But is this equation always true? What if $x = -9$ and $y = -25$. Let's evaluate $\sqrt{x \times y}$ and $\sqrt{x} \times \sqrt{y}$ again.

$$\sqrt{x \times y} = \sqrt{(-9) \times (-25)} = \sqrt{225} = 15$$

$$\text{and} \quad \sqrt{x} \times \sqrt{y} = \sqrt{(-9)} \times \sqrt{(-25)}$$

but here we run into a difficulty because, in the world of real numbers, the square root of a negative number does not exist. However, we can get around the problem by writing -9 as -1×9 and -25 as -1×25 so that now:

$$\sqrt{x} \times \sqrt{y} = \sqrt{(-1 \times 9)} \times \sqrt{(-1 \times 25)}$$

which according to the relationship $\sqrt{x \times y} = \sqrt{x} \times \sqrt{y}$ can be expanded to form:

$$\sqrt{x} \times \sqrt{y} = \sqrt{(-1)} \times \sqrt{(9)} \times \sqrt{(-1)} \times \sqrt{(25)}$$

and can be reordered as:

$$= \sqrt{(-1)} \times \sqrt{(-1)} \times \sqrt{(9)} \times \sqrt{(25)}$$

$$= \left(\sqrt{(-1)}\right)^2 \times 3 \times 5$$

Since, for any quantity, z, $(\sqrt{(z)})^2 = z$, then $\left(\sqrt{(-1)}\right)^2 = -1$.

Upon substituting into the equation above, this leads to:

$$\sqrt{x} \times \sqrt{y} = -1 \times 3 \times 5 = -15$$

Therefore, in this case, $\sqrt{x \times y} \neq \sqrt{x} \times \sqrt{y}$.

In fact, $\sqrt{x \times y} = \sqrt{x} \times \sqrt{y}$ is true for all cases, ***except*** where both x and y have negative values.

Quiz

1. What can you say about a and b in this equation; $\sqrt{a} \times \sqrt{b} = -\sqrt{a \times b}$?

2. Simplify $\sqrt{-9x^2} \times \sqrt{-4y^4}$.
3. What is the product of the square roots of −49 and −81?

1. *a* and *b* both have negative values.
2. $-6xy^2$
3. -63

Square root of a negative number

As we declared in the last section, in the world of real numbers the square root of a negative number does not exist. This is simply because the square of a positive number is a positive number and the square of a negative number is also a positive number. For example, $3^2 = 9$ and $(-3)^2 = 9$. Notice the importance of the brackets here as they indicate that it is the −3 which is being multiplied by itself. Without them $-3^2 = -9$ because the order of precedence of operations here requires that the 3 is raised to the power of 2 before the negative sign is applied. You could try $(-3)^2$ and -3^2 in your calculator and you should get 9 and −9, respectively.

This indicates that the squares of all real numbers are positive. Therefore, the square root of a negative but real number cannot be a real number. But not all numbers are real; there exists a branch of mathematics that deals with **complex numbers** which have real and imaginary parts. A complex number is of the form of $a + ib$ where a and b are real numbers and $i = \sqrt{-1}$. Thus, a is the real part and ib is the imaginary part of the complex number $a + ib$. So $i^2 = \left(\sqrt{-1}\right)^2 = -1$.

$$\text{Therefore,} \quad \sqrt{-x} = \sqrt{-1}\sqrt{x} = i\sqrt{x}.$$

Some interesting complex number relationships follow from this definition. For example:

$$\text{Addition}\,(a + ib) + (c + id) = (a + c) + i\,(b + d)$$

$$\text{Multiplication } (a+ib)\times(c+id) = ac + iad + ibc + i^2bd$$
$$= (ac-bd) + i(ad+bc)$$
$$(a+ib)^2 = a^2 + 2iab + i^2b^2 = \left(a^2-b^2\right) + 2iab$$
$$(a-ib)^2 = a^2 - 2iab + i^2b^2 = \left(a^2-b^2\right) - 2iab$$
$$(a+ib)(a-ib) = a^2 - i^2b^2 = a^2 + b^2$$

Quiz

Which of the following relationships are true?

1. $i^3 = -i$
2. $i^4 = 1$
3. $i^5 = i$
4. $(a+ib)^3 = a(a^2-3b^2) + ib(3a^2-b^2)$
5. $(a+ib)^4 = (a^4 - 6a^2b^2 + b^4) + 4iab(a^2-b^2)$
6. $\dfrac{(a+ib)}{(a-ib)} = \left(\dfrac{a^2-b^2}{a^2+b^2}\right) + i\left(\dfrac{2ab}{a^2+b^2}\right)$

They are all true. Here are some of the intermediate steps:

4. $(a+ib)^3 = a^3 + 3ia^2b + 3i^2ab^2 + i^3b^3 = a(a^2-3b^2) + ib(3a^2-b^2)$
5. $(a+ib)^4 = a^4 + 4ia^3b + 6i^2a^2b^2 + 4i^3ab^3 + i^4b^4$
$= a^4 + 4ia^3b - 6a^2b^2 - 4iab^3 + b^4$
$= (a^4 - 6a^2b^2 + b^4) + 4iab(a^2-b^2)$
6. $\dfrac{(a+ib)}{(a-ib)} = \dfrac{(a+ib)(a+ib)}{(a-ib)(a+ib)} = \dfrac{(a^2-b^2)+2iab}{a^2+b^2} = \left(\dfrac{a^2-b^2}{a^2+b^2}\right) + i\left(\dfrac{2ab}{a^2+b^2}\right)$

Power or root of a product

Can the power of a product be expressed like this?

$$(ax)^b = ax^b$$

where b is the power that the product of a and x is raised to. Superficially this may look correct, but to expand the brackets in this way is to misunderstand the meaning of the brackets. The expression $(ax)^b$ indicates that it is the product of a and x that is raised to the power of b, not just x. The rule is that the power of a product of variables is equal to the product of the individual variables each raised to the power.

Therefore, $(ax)^b = a^b x^b$, not ax^b as suggested above.

A simple numerical example will show that the equation at the top of this page cannot be correct. If $a = 3$, $x = 4$ and $b = 2$ then:

$$(ax)^b = (3 \times 4)^2 = 12^2 = 144$$

but $ax^b = 3 \times 4^2 = 3 \times 16 = 48$

however $a^b x^b = (3 \times 4)^2 = 3^2 \times 4^2 = 9 \times 16 = 144.$

Similarly:

$$\sqrt[b]{ax} = ax^{\frac{1}{b}}$$

cannot be true. The expression $\sqrt[b]{ax}$ indicates that this is the bth root of the product ax. The bth root is an alternative notation for raising to the power of the reciprocal of b. So, for example, $\sqrt[b]{x} = x^{\frac{1}{b}}$. Therefore:

$$\sqrt[b]{ax} = (ax)^{\frac{1}{b}}$$

So, from above:

$$(ax)^{\frac{1}{b}} = a^{\frac{1}{b}} x^{\frac{1}{b}} \text{ not } ax^{\frac{1}{b}} \text{ as suggested above.}$$

The following numerical example shows that $\sqrt[b]{ax} = ax^{\frac{1}{b}}$ cannot be correct. If $a = 8$, $x = 27$ and $b = 3$, then:

$$\sqrt[b]{ax} = \sqrt[3]{8 \times 27} = \sqrt[3]{216} = 6$$

$$\text{but} \quad ax^{\frac{1}{b}} = 8 \times 27^{\frac{1}{3}} = 8 \times \sqrt[3]{27} = 8 \times 3 = 24$$

$$\text{however} \quad a^{\frac{1}{b}} x^{\frac{1}{b}} = 8^{\frac{1}{3}} 27^{\frac{1}{3}} = \sqrt[3]{8} \times \sqrt[3]{27} = 2 \times 3 = 6.$$

Quiz

Which is the correct answer for each of the following?

1.	$3^2 \times (2^3 \times 3)^3 =$	a. $2^9 \times 3^5$	b. $2^6 \times 3^5$	c. $2^6 \times 3^3$
2.	$2^4 \times (2 \times 3^3)^3 =$	a. $2^5 \times 3^9$	b. $2^7 \times 3^9$	c. $2^5 \times 3^6$
3.	$\sqrt[3]{x^6 y^9} =$	a. $x^2 y^3$	b. $x^3 y^6$	c. $x^3 y^2$
4.	$(a^2 b^8)^{\frac{1}{4}} =$	a. $a\sqrt{b}$	b. $\sqrt{ab^2}$	c. $\sqrt{a}\, b^2$
5.	$\left(\sqrt[3]{125} \times \sqrt[4]{256}\right)^2 =$	a. 240	b. 400	c. 100

1. a 2. b 3. a 4. c 5. b

Expanding brackets

Why is this not true; $(x + y)^2 = x^2 + y^2$?

The answer is in the meaning of the brackets. The brackets mean that their contents must be treated as a whole, not as individual parts.

If we had the expression $2(a + b)$, that would mean two multiplied by the sum of a and b. This is the same as the sum of the product of 2 and a and the product of 2 and b. In the form of an equation this is:

$$2(a+b)=2a+2b$$

This is fairly basic stuff but it is an important point to establish. So $(x+y)^2$ is the product of $(x+y)$ and $(x+y)$;

$$(x+y)^2=(x+y)\times(x+y).$$

In other words, all that is in the first bracket must be multiplied by all that is in the second bracket. There are various tricks and methods for making sure that nothing is missed out, or included twice, when evaluating such a product. Here are some.

Product by areas

Think of $(x + y)$ as a linear measure. In that case, $(x + y)^2$ will represent an area with sides of $(x+y)$ and $(x+y)$, which could be drawn like this.

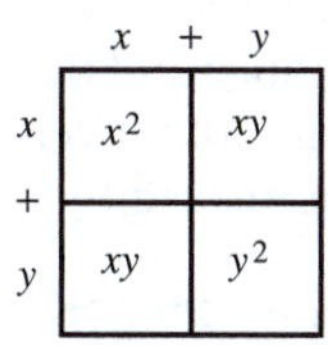

The complete area is made up of four parts: x^2, xy, xy and y^2. Therefore:

$$(x+y)^2=x^2+2xy+y^2$$

For example, if $x=3$ and $y=4$ then:

$$7^2=(3+4)^2=3^2+2\times3\times4+4^2=9+24+16=49$$

Smiley face

Draw curved lines above and below the product of $(x+y)$ and $(x+y)$ as shown. Each line connects a term in the first pair of brackets with a term in the second pair of brackets. Think of the equation as the mid-part of a face and the curved lines above as (overlapping) eyebrows and the curved lines below as a pair of lips. In that case, the products of terms described by the anatomy are: left eyebrow x^2, right eyebrow y^2, top lip yx and bottom lip xy. The sum of these is $x^2+2xy+y^2$.

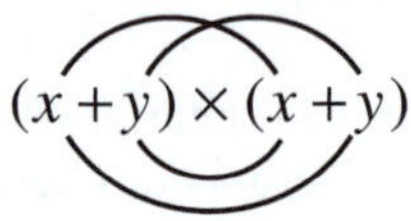

Eagle and tortoise

Some might consider the last method as a bit 'junior'. In that case, here is another variation on the same theme. The ancient Greek playwright Aeschylus, it is said, was killed when a passing eagle decided to crack open the tortoise it was carrying by dropping it onto his bald head, mistaking it for a rock. Here is a picture of an eagle carrying a tortoise. Now the curved lines above the expression are the eagle's wings and the lines below the expression are the eagle's beak carrying the inverted tortoise. Again, this is the sum of the products connected by the lines: $x^2+2xy+y^2$.

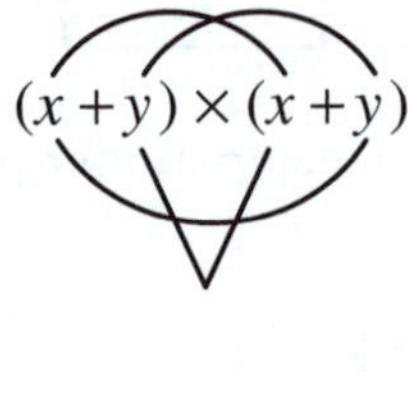

FOIL

$(x+y)\times(x+y)$ Foil stands for First, Outer, Inner and Last which relate to the products of individual terms in the overall product.

So the products of the:

First terms of each bracket is x^2
Outer, that is, 1st term of the 1st bracket and the 2nd term of the 2nd bracket, is xy
Inner, that is, 2nd term of the 1st bracket and the 1st term of the 2nd bracket, is yx
Last terms of each bracket is y^2

And once again the sum of the products is $x^2 + 2xy + y^2$.

Quiz

Can you find the following products?

1. $(p+q) \times (p+q)$
2. $(p-q) \times (p-q)$
3. $(p+q) \times (p-q)$
4. $(100 + 0.01) \times (100 + 0.01)$ exactly, but without using a calculator.
5. $(4\frac{1}{8})^2$ exactly, as another mixed number, but without using a calculator. (A mixed number is a combination of a whole number and a proper fraction, for example, as in this question.)
6. Without using a calculator, can you find the area of a rectangular carpet that measures 9.5 metres by 10.5 metres?

1. $p^2 + 2pq + q^2$
2. $p^2 - 2pq + q^2$
3. $p^2 - q^2$
4. 10 002.0001
5. $17\frac{1}{64}$
6. 99.75 m^2

Powers of negative values

Some people get rather confused when raising negative quantities to powers and may be tempted to write $(-x)^2 = -x^2$.

This reflects a misunderstanding of the meaning of brackets because $(-x)^2$ means $-x$ multiplied by itself two times, that is $(-x) \times (-x)$. In other words, it is not just the x terms that are multiplied together but also the negative indicators; the minus signs. So two negative terms multiplied together give a positive product. Three negative terms multiplied together give a negative product. Four negative terms multiplied together give a positive product, and so on. We could express this as:

$$(-x)^2 = x^2 \qquad (-x)^3 = -x^3$$

$$(-x)^4 = x^4 \qquad (-x)^5 = -x^5, \text{ etc.}$$

In fact, $(-x)^n = x^n$ if n is even and $(-x)^n = -x^n$ if n is odd.

Quiz

Can you find the real solutions for the following expressions without using a calculator?

1. $(-1)^{317}$
2. $(-1)^8$
3. $\sqrt[3]{-1}$
4. $\sqrt[97]{-1}$
5. $\sqrt[4]{-1}$
6. $(-1)^{235} + (-1)^{68}$

1. -1 2. 1 3. -1 4. -1 5. No real solution.
6. 0

The root sign

What does the symbol $\sqrt{\ }$ actually mean? We recognise this as the square root sign such that if $x = \sqrt{y}$ then, upon squaring both sides of the equation, $y = x^2$. However, since square and square root are inverse functions of each other, that implies that $\sqrt{x^2} = x$, which it is. But it is not the only solution; the other is $\sqrt{x^2} = -x$. The two solutions may be written as $\sqrt{x^2} = \pm x$, meaning that the solution of $\sqrt{x^2}$ may take the positive or negative value of x. To distinguish between the two solutions we call the positive solution the principal solution. In many mathematical applications only the principal solution is permitted, but we may recognise the other solution by using the $\pm$ symbol as we do, for example, in the formula solution for quadratic equations.

We have to be careful how we use these solutions. If we were to say that $\sqrt{x^2} = x$ and that $\sqrt{x^2} = -x$ we could then deduce that $x = -x$, which of course is ridiculous. For example, the two solutions of $x^2 = 4$ are $x = 2$ and $x = -2$, but $2 \neq -2$; the two solutions are true, but not simultaneously. So if there are two real solutions for y when $y = \sqrt{x}$, provided that x is positive, how many real solutions for y are there when $y = \sqrt[4]{x}$? Disappointingly, there are still only two. For example, if $y = \sqrt[4]{16}$ the solutions are $y = 2$ and $y = -2$.

On the other hand, if $y = \sqrt[3]{x}$ there is always one real solution for y irrespective of the sign of x. For example, when $y = \sqrt[3]{8}$ then $y = 2$, but if $y = \sqrt[3]{-8}$ then $y = -2$; there are no other real solutions.

So if $y = \sqrt[n]{x}$ there is always one real solution for y if n is an odd integer. If n is an even integer there are two real solutions for y, but only if x is positive; otherwise there are none.

How does this all affect non-integer roots? For example, if we want to find $\sqrt[n]{x}$ where n is not an integer it all depends whether x is positive or negative and whether n is rational or irrational. A rational number is one that can be expressed as a fraction, so for example $\frac{2}{3}$ and $\frac{857}{271}$ are rational but π and $\sqrt{2}$ are irrational.

If n is irrational, only one real solution exists for $\sqrt[n]{x}$ and only then provided that x is positive. The solution is always positive. For example, $\sqrt[\pi]{10} \approx 2.081$ but $\sqrt[\pi]{-10}$ has no real solution.

If n is rational, then the availability of real solutions depends upon whether the denominator of n is odd or even. If n can be written as $n = \frac{p}{q}$, where p and q are whole numbers and $y = \sqrt[n]{x}$, then $y = \sqrt[\frac{p}{q}]{x}$. This may be expressed as $y = x^{\frac{q}{p}}$ because $\sqrt[n]{x} = x^{\frac{1}{n}}$, from the meaning of reciprocal powers. Then $y = x^{\frac{q}{p}} = (\sqrt[p]{x})^q$, which comes from the power of a power rule for indices.

In the expression $(\sqrt[p]{x})^q$, if p is even and x is positive, there are two real solutions because q must be odd. Both the whole numbers p and q cannot be even because if they were they could both be divided by a common factor of 2.

For example, if the two solutions are allowed, $\left(\sqrt[4]{16}\right)^3 = \pm 8$.

If p is odd, there is only one real solution irrespective of the sign of x and the value of q. For example, $\left(\sqrt[5]{243}\right)^3 = 27$, $(\sqrt[5]{-243})^3 = -27$, $\left(\sqrt[5]{243}\right)^2 = 9$ and $(\sqrt[5]{-243})^2 = 9$.

If q is odd but p is even, there is no real solution when x is negative but there are two real solutions when x is positive. For example, $\left(\sqrt[4]{81}\right)^3 = \pm 27$ but $(\sqrt[4]{-81})^3$ has no real solution.

Quiz

Does each of the following have the full correct answer? If not, can you find the right answer?

1. $32^{\frac{2}{5}} = 4$
2. $32^{\frac{3}{5}} = 8$
3. $(-32)^{\frac{2}{5}} = -4$
4. $(-32)^{\frac{3}{5}} = -8$
5. $16^{\frac{3}{4}} = 8$
6. $16^{\frac{5}{4}} = -32$
7. $(-16)^{\frac{3}{4}} = -8$

1. Yes. 2. Yes. 3. No, $(-32)^{\frac{2}{5}} = 4$. 4. Yes. 5. No, $16^{\frac{3}{4}} = \pm 8$. 6. No, $16^{\frac{5}{4}} = \pm 32$. 7. No, $(-16)^{\frac{3}{4}}$ has no real solution.

Splitting the numerator

Sometimes it is desirable to split one fraction into the sum of two other fractions, for example, to simplify a fractional expression. So, is this the way to do it; $\frac{x^2+y}{x} = x + y$? An answer presented like this would indicate that there is a misunderstanding of the nature of fractions and how to cancel within them.

A fraction consists of a numerator and a denominator; the numerator is on the top and the denominator is on the bottom.

$$\frac{\text{Numerator}}{\text{Denominator}}$$

The meaning is that the numerator is divided by the denominator; that is to say, it is the whole of the numerator that is divided by the whole of the denominator. It does not change the value of the fraction if you multiply or divide both numerator and denominator by the same quantity. For example, if you multiply the numerator and the denominator of $\frac{1}{2}$ by 2 you get $\frac{1\times 2}{2\times 2} = \frac{2}{4}$; half a pizza is the same as two quarters.

So, to take a simple numerical example, if we had the fraction $\frac{4+6}{2}$ then we could first add the two terms in the numerator to give $\frac{10}{2}$ and divide both numerator and denominator by 2 to get $\frac{5}{1} = 5$; this last step is what we normally understand by cancelling.

An alternative route to the answer would be to start with $\frac{4+6}{2}$ and then divide the whole of the numerator and the whole of the denominator by 2. That leads to $\frac{2+3}{1} = \frac{5}{1} = 5$; as you would expect, this gives the same answer again.

Now let's consider the original expression, $\frac{x^2+y}{x}$, which, upon dividing the whole of the numerator and the whole of the denominator by x, becomes $\frac{x+y/x}{1} = x + \frac{y}{x}$.

So, if the numerator is a sum of terms, then each term has to be divided by the whole of the denominator.

Quiz

Which of the following are correct? For those which are not correct, can you find the right answer?

1. $\dfrac{9+15}{3} = 8$
2. $\dfrac{x^3+xy^2}{x^2} = x + \dfrac{y^2}{x}$
3. $\dfrac{ab^2c^3+a^3b^2c}{abc} = b^2(c+a)$
4. $\dfrac{2pq-3qr+5pr}{30pqr} = \dfrac{1}{15r} - \dfrac{1}{10q} + \dfrac{1}{6r}$
5. $\dfrac{2ab+4bc-3cd}{abc} = \dfrac{2}{c} + \dfrac{4}{a} - \dfrac{3}{b}$
6. $\dfrac{5j+4jk+3jkl}{jl} = 5j + \dfrac{4k}{l} + 3kl$

1. Correct.
2. Correct.
3. $\dfrac{ab^2c^3+a^3b^2c}{abc} = bc^2 + a^2b = b(c^2+a^2)$
4. $\dfrac{2pq-3qr+5pr}{30pqr} = \dfrac{1}{15r} - \dfrac{1}{10p} + \dfrac{1}{6q}$

5. $\dfrac{2ab+4bc-3cd}{abc} = \dfrac{2}{c} + \dfrac{4}{a} - \dfrac{3d}{ab}$

6. $\dfrac{j!}{5j!+4j!k+3j!kl} = \dfrac{1}{5} + \dfrac{1}{4k} + 3k = \dfrac{1}{5+4k} + 3k$

Splitting the denominator

In the last section, we saw how one fraction could be represented by two fractions, each with the same denominator as the original fraction and with numerators which were the separate terms of the numerator of the original fraction. For example, $\frac{x+y}{z} = \frac{x}{z} + \frac{y}{z}$. But what if it is the denominator that is a sum of terms? How can we split $\frac{z}{x+y}$? Surely as $\frac{z}{x} + \frac{z}{y}$.

Again, this would be a significant misunderstanding of the nature of fractions and how to manipulate them. Whereas a function can be split when the numerator is a sum, it cannot be split if the sum is only in the denominator. A simple numerical example will show that it cannot be done. If $z = 10$, $x = 2$ and $y = 5$, then:

$$\frac{z}{x+y} = \frac{10}{2+5} = \frac{10}{7}$$

$$\text{but} \quad \frac{z}{x} + \frac{z}{y} = \frac{10}{2} + \frac{10}{5} = 5 + 2 = 7$$

Therefore, $\frac{z}{x+y} \neq \frac{z}{x} + \frac{z}{y}$. Actually, there is nothing meaningful we can do to split or simplify $\frac{z}{x+y}$.

Quiz

Which of the following are correct? For those that are not correct, what is the right answer?

1. $\dfrac{14}{2+5} = 2$

2. $$\frac{x}{x^2+x^3} = \frac{1}{x} + \frac{1}{x^2}$$
3. $$\frac{ab}{a+b} = \frac{ab}{a} + \frac{ab}{b} = b + a$$
4. $$\frac{ab+bc}{a+b+c} = \frac{ab}{a} + \frac{1}{b} + \frac{bc}{c} = b + \frac{1}{b} + b = 2b + \frac{1}{b}$$
5. $$\frac{xy}{yx+xy} = \frac{xy}{yx} + \frac{xy}{xy} = 1 + 1 = 2$$
6. $$\frac{6\pi}{12r^2+3p} = \frac{6\pi}{15r^2p} = \frac{2\pi}{5r^2p}$$

1. Correct.
2. $$\frac{x}{x^2+x^3} = \frac{1}{x+x^2}$$
3. $$\frac{ab}{a+b}$$ no simplification possible
4. $$\frac{ab+bc}{a+b+c} = \frac{b(a+c)}{a+b+c}$$
5. $$\frac{xy}{yx+xy} = \frac{1}{1+1} = \frac{1}{2}$$
6. $$\frac{6\pi}{12r^2+3p} = \frac{2\pi}{4r^2+p}$$

Assuming unique solutions

The solution of the equation $x^2 = bx$ may be found by dividing both sides by x, which leads to $x = b$; true or false?

The answer is both. It is true that $x = b$ is one solution to the equation $x^2 = bx$, but it is not the only one; the other is $x = 0$. So how should we have known that there were two solutions and indeed how could we find them?

The equation $x^2 = bx$, written in the form $x^2 - bx = 0$, has a polynomial expression in x on its left-hand side. This means that it is an expression in a given number of terms and is made up of the variable x raised only to positive whole-number powers, coefficients and the operators of addition and subtraction.

The maximum number of independent solutions for an equation that contains only a polynomial expression is equal to the numerical value of the highest power of x, 2 in this case. The solutions of a polynomial equation do not necessarily have to be all different and they do not have to be real.

Where possible, the easiest solution for a polynomial equation which equals zero is found by finding the factors of the polynomial. In this case, our equation could be written as $x(x-b)=0$ and the factors are x and $(x-b)$. Since their product is zero then either $x=0$ or $x=b$, and so these are the two solutions.

A polynomial equation with a maximum power of 2 is called a quadratic and can be written in the general form of $ax^2+bx+c=0$. If the expression on the left-hand side cannot easily be factorised, then a solution may always be found by using the quadratic equation:

$$x=\frac{-b\pm\sqrt{b^2-4ac}}{2a}$$

There are two solutions because the $\pm$ operator indicates that it may be $+$ or $-$.

If $b^2-4ac>0$, there are two distinct real solutions as just described. If $b^2-4ac=0$, then both solutions are the same and therefore there is only one distinct real solution. If $b^2-4ac<0$, that calls for the square root of a negative number and so there are no real solutions; they are both complex numbers. So the solutions of the following equations are:

$$3x^2+2x-33=0$$

$$x=\frac{-2\pm\sqrt{2^2-4\times3\times(-33)}}{2\times3}=\frac{-2\pm\sqrt{400}}{6}=\frac{-2\pm20}{6},$$

$$\text{so } x=3 \text{ or } x=-\frac{11}{3}$$

There are two distinct real solutions because $b^2 - 4ac > 0$; $b^2 - 4ac = 400$.

$$x^2 + 14x + 49 = 0$$

$$x = \frac{-14 \pm \sqrt{14^2 - 4 \times 1 \times 49}}{2 \times 1} = \frac{-14 \pm \sqrt{0}}{2} = \frac{-14}{2},$$

so $x = -7$

There is only one distinct real solution because $b^2 - 4ac = 0$.

$$2x^2 + 3x + 7 = 0$$

$$x = \frac{-3 \pm \sqrt{3^2 - 4 \times 2 \times 7}}{2 \times 2} = \frac{-3 \pm \sqrt{-47}}{4}$$

There are no real solutions because $b^2 - 4ac < 0$; $b^2 - 4ac = -47$.

Quiz

1. Without working it out, suggest the maximum possible number of solutions for the equation $x^4 - 8x^3 + 21x^2 - 26x + 24 = 0$.

   How many real and distinct solutions are there for the following equations?
2. $16x^2 + 8x + 1 = 0$
3. $x^2 + 16x + 4 = 0$
4. $x^2 + x + 4 = 0$

1. 4
2. 1 real and distinct solution.
3. 2 real and distinct solutions.
4. 0 real and distinct solutions.

Interpreting factors

Consider this; the solution of the equation $x^2+3x+2=0$ is that x can take the value of 1 or 2 because the factors of the expression are $x+1$ and $x+2$.

What has gone amiss here? This is a quadratic equation and the values for x can be determined by factorising the original expression x^2+3x+2. The factors are indeed $x+1$ and $x+2$. Notice that the original equation was set up to equal zero. Therefore, the product of the factors must also equal zero, that is,

$$(x+1)(x+2)=0$$

The only way this can be true is if either:

$$x+1=0 \quad \text{or} \quad x+2=0$$

This is because the product of any two non-zero quantities must also be a non-zero quantity. Now the solution to the equation, implied by the factors, is that $x=-1$ or $x=-2$. Notice that the magnitude of the values of x are the same as those incorrectly derived above but their signs are opposite.

Quiz

Which of the following are correct?

1. $x^2+x-2=(x-1)(x+2)=0$ so $x=1$ or $x=-2$
2. $2x^2-5x-3=(x-3)(2x+1)=0$ so $x=\dfrac{1}{2}$ or $x=-3$
3. $x^3-2x^2-5x+6=(x-1)(x+2)(x-3)=0$ so $x=-1$ or $x=2$ or $x=-3$

1. Correct.
2. No, $x=-\dfrac{1}{2}$ or $x=3$
3. No, $x=1$ or $x=-2$ or $x=3$

Misunderstanding square root signs

What is wrong here: $\frac{1}{\sqrt{3a+2}+\sqrt{3a+2}} = \frac{1}{3a+2}$?

Some people get a little confused by the meaning of the square root sign. It indicates that it is the value which when multiplied by itself gives the term under the square root sign. A square root is the inverse function of a square. Therefore, the square of the square root of a term is the same term, $\left(\sqrt{x}\right)^2 = x$. Likewise, the square root of the square of a term is also the same term, $\sqrt{x^2} = x$.

The problem with the equation above is that the writer has confused multiplication with addition. Had the problem been to simplify $\frac{1}{\sqrt{3a+2}\times\sqrt{3a+2}}$ that could be written as:

$$\frac{1}{\sqrt{3a+2}\times\sqrt{3a+2}} = \frac{1}{\left(\sqrt{3a+2}\right)^2} = \frac{1}{3a+2}$$

However, the problem as written can only be simplified as:

$$\frac{1}{\sqrt{3a+2}+\sqrt{3a+2}} = \frac{1}{2\times\sqrt{3a+2}}$$

What is wrong here: $\frac{\sqrt{3+b}-\sqrt{3+2a+b}}{a} = \frac{\sqrt{3+b}-\sqrt{3}-2a-b}{a} = \frac{-2a}{a} = -2$?

Again, the meaning of the square root sign has been confused. It is the square root of the whole of the expression under the whole of the square root sign that is required, not just the first or any other part of it. There is no meaningful simplification of $\frac{\sqrt{3+b}-\sqrt{3+2a+b}}{a}$.

Quiz

Which of the following are correct?

1. $\sqrt{x^2} = \left(\sqrt{x}\right)^2$
2. $\sqrt{\left(\sqrt{a^2}\right)^2} = a$

3. $\sqrt{\sqrt{a^4}} = a$
4. $\sqrt{a^2 + a^2} = 2a$
5. $\sqrt{2a^2 + 2a^2} = 2a$
6. $\sqrt{3+b} \times \sqrt{3+2a+b} = \sqrt{9+6a+6b+2ab+b^2}$

1. Correct.
2. Correct.
3. Correct.
4. No, $\sqrt{a^2 + a^2} = \sqrt{2}a$.
5. Correct.
6. Correct.

Functions

If $f(a) = \sqrt{3a+5}$ then $f(a+b) = \sqrt{3a+5} + b$; true or not?

Functions are often referred to as function engines; black boxes where something goes in, is acted upon, and something else comes out. If your function is $f(a) = a^2$ that means that the input is a and the output is a^2 so, in this case, the function takes the input and squares it to give the output.

For the problem above the function is of a, $f(a) = \sqrt{3a+5}$, so the function takes the input, multiplies it by 3, then adds 5, and then takes the square root of the result. So if we require the same function, but this time of $a+b$, then we follow the same process; we multiply the input by 3, add 5, and take the square root of the result but this time with an input of $a+b$. In this case:

Input	$a+b$
Multiply by 3	$3a+3b$
Add 5	$3a+3b+5$
Take square root	$\sqrt{3a+3b+5}$

Therefore, $f(a+b)=\sqrt{3a+3b+5}$. $f(a+b)\neq f(a)+b$ as was suggested in the opening line above.

Is this true? If $f(b)=-b$ and $g(a)=a^2$ then $f(g(a))=-a^2b$?

This is all about functions of functions. $f(g(a))$ means take the output of function engine $g(a)$ and use it as input to function engine $f(b)$. If a is the input to $g(a)$ then, as $g(a)$ outputs the square of the input, the output is a^2. $f(b)$ outputs the input multiplied by -1. So if the input to $f(b)$ is a^2 then its output must be $-a^2$, that is $f(g(a))=-a^2$.

Notice that $f(g(a))\neq g(f(a))$. In this example $f(b)=-b$, therefore $f(a)=-a$. So $g(f(a))$ means: take the output of $f(a)$, which is $-a$, and use this as the input to $g(a)$ which outputs the square of the input. Therefore, $g(f(a))=(-a)^2=a^2$.

Quiz

Which of the following are correct? If any are not correct, what are their right answers?

1. If $f(a)=a+3$ then $f(2a+5)=2a+5+3=2a+8$.
2. If $f(x)=x^2-2$ then $f(x^2+1)=\left(x^2+1\right)^2-2=x^4+2x^2-1$.
3. If $f(a)=\sqrt{a-1}$ then $f(3a^2+1)=3\left(\sqrt{a-1}\right)^2+1=3a^2$.
4. If $f(b)=\sqrt[3]{b^2-2b}$ then $f(b^3+1)=\left(b^2+2b\right)^3+1=b^6+6b^5+12b^4+8b^3+1$.
5. If $f(x)=-x^2$ and $g(y)=y+y^2$ then $f(g(z))=-\left(z+z^2\right)^2=-z^2(1+z)^2$.
6. If $f(x)=-x^2$ and $g(y)=y+y^2$ then $g(f(z))=-\left(z+z^2\right)^2=-z^2(1+z)^2$.

1. Correct.
2. Correct.
3. No, $f(3a^2+1)=\sqrt{3a^2+1-1}=\sqrt{3}\,a$.

4. No, $f(b^3+1) = \sqrt[3]{(b^3+1)^2 - 2(b^3+1)} = \sqrt[3]{b^6+2b^3+1-2b^3-2} = \sqrt[3]{b^6-1}$.
5. Correct.
6. No, $g(f(z)) = -z^2 + (-z^2)^2 = -z^2 + z^4 = z^2(z^2-1)$.

'Multiplying through' confusion

If we want to make x the subject of this equation $\frac{1}{\pi^2} + 2x = \pi^2$, we will need to multiply through by π^2 and the result will then be $x = \frac{1}{2}\pi^4$; true or not?

'Multiplying through' is often a convenient way to simplify an equation that involves fractions into one that does not. With any equation, you can do whatever you like to it, provided that you do the same to the whole of both sides of the equation. Think of an equation as a set of scales. The expression on the left-hand side of the equation sits in the left-hand pan of the scales and the expression on the right-hand side sits in the right-hand pan. The equals sign only applies if the scales balance. If the scales do balance, then adding or subtracting the same amount to or from both pans will not affect the balance. Likewise, multiplying or dividing the contents of both pans by the same amount also will not affect the balance.

So, for example, if an equation is $3+5=8$ then adding 4 to both sides does not affect the balance because $3+5+4=8+4$. If we multiply the whole of both sides by 6 then $6\times3+6\times5=6\times8$ and the balance is maintained. Other operations such as taking the square, the cube root or the logarithm of the whole of both sides also do not affect the balance; $(3+5)^2=8^2$, $\sqrt[3]{3+5}=\sqrt[3]{8}$ and $\log_{10}(3+5)=\log_{10}8$.

However, if you do not apply your chosen operation to the whole of both sides, but only to part, the equation is no longer true. For example, $3^2+5\neq8^2$, $\sqrt[3]{3}+5\neq\sqrt[3]{8}$ and $\log_{10}(3)+5\neq\log_{10}8$. You can check any of these with a calculator.

So, let's go back to the problem at the beginning of this section. If we wish to make x the subject of the equation $\frac{1}{\pi^2} + 2x = \pi^2$, then an order of operations could be:

multiply the whole of both sides by π^2	$\pi^2 \frac{1}{\pi^2} + 2x\pi^2 = \pi^2 \times \pi^2$
hence	$1 + 2x\pi^2 = \pi^4$
subtract 1 from both sides	$1 + 2x\pi^2 - 1 = \pi^4 - 1$
hence	$2x\pi^2 = \pi^4 - 1$
divide the whole of both sides by $2\pi^2$	$\frac{2x\pi^2}{2\pi^2} = \frac{\pi^4 - 1}{2\pi^2}$
hence	$x = \frac{\pi^4 - 1}{2\pi^2}$.

Quiz

'Multiplying through' to eliminate the fractions in the following equations has taken place in questions 1 to 3 below. Which have been done correctly?

1. $\frac{7}{a^2} - \frac{12}{a} = 5$ becomes $7 - 12a = 5a^2$.
2. $\frac{b}{5} - \frac{5}{b} = 5$ becomes $b - 25 = 25b$.
3. $\frac{a^2}{b^3} + \frac{b^2}{a^3} = 1$ becomes $a^6 + b^6 = a^3b^3$.
4. Find x to the nearest whole number from $17 - \frac{x}{12.34^2} = \pi$.
5. Find y to 2 decimal places from $\frac{y}{47} - \frac{23}{48} = \frac{24}{49}$.
6. Find z to 2 decimal places from $\frac{z^2}{3} - \frac{24}{z} = 0$.

1. Correct.

2. No, multiply both sides of $\frac{b}{5} - \frac{5}{b} = 5$ by $5b$ to get $b^2 - 25 = 25b$.
3. No, multiply both sides of $\frac{a^2}{b^3} + \frac{b^2}{a^3} = 1$ by a^3b^3 to get $a^5 + b^5 = a^3b^3$.
4. 2110
5. 45.54
6. 4.16

Inequalities

If $\frac{1}{x} > 123$ then $x > \frac{1}{123}$?

An expression of inequality shows the relationship between terms that are not the same. If we return to the analogy of using a balanced set of scales to represent equality, then inequality is represented by an unbalanced set of scales; one side is heavier (greater) than the other.

The main inequality signs mean:

$a \neq b$	a is not the same as b, but it does not say which is the greater.
$a > b$	a is greater than b.
$a < b$	a is less than b.
$a \geq b$	a is greater than or equal to b.
$a \leq b$	a is less than or equal to b.

So what can we say about inequalities?

If $a > b$ and $b > c$, then $a > c$. For example, $5 > 3$ and $3 > 2$, therefore $5 > 2$.

If we add or subtract the same amount to both sides of the inequality, the inequality still holds. For example, if $a > b$ then $a + c > b + c$ so, since $5 > 3$, then $5 + 2 > 3 + 2$.

If we multiply or divide both sides of the inequality by the same positive amount, the inequality still holds. For example, if $a > b$ then, if $c > 0$, $\frac{a}{c} > \frac{b}{c}$. It is true that $5 > 3$, so $\frac{5}{2} > \frac{3}{2}$.

However, if we multiply or divide both sides of the inequality by the same negative amount, the inequality reverses. For example, if $a > b$ and $c < 0$, then $\frac{a}{c} < \frac{b}{c}$. It is true that $5 > 3$, but $\frac{5}{-2} < \frac{3}{-2}$, which is the same as $-2.5 < -1.5$.

If we apply the reciprocal to both sides of the inequality, the inequality reverses. For example, if $a > b$ then $\frac{1}{a} < \frac{1}{b}$. It is true that $5 > 3$, but $\frac{1}{5} < \frac{1}{3}$.

Now let's go back to the inequality at the beginning of this section. From the last paragraph the inequality must be wrong. The correct statement would be that if $\frac{1}{x} > 123$ then $x < \frac{1}{123}$.

Quiz

Which of the following inequality statements are true?

1. If $a < b$ and $b < c$ then $c > a$.
2. If $\dfrac{x}{y} > \dfrac{y}{x}$ then $x > y$.
3. If $p^{-1} > q$ then $p > q^{-1}$.
4. If $s > t$ then $-\dfrac{3}{s} > \dfrac{-3}{t}$.
5. If $g^3 > h^3$ then $g > h$ for any values of g and h.
6. If $m > n$ then $99 - m > 99 - n$.

1. True.
2. True.
3. No, if $p^{-1} > q$ then $p < q^{-1}$ (the reciprocal rule).
4. True, both the reciprocal rule and the multiplication by a negative number rule apply.

5. True, even if h is negative or g and h are negative.
6. No, multiplying both sides by -1 changes the inequality even though the subsequent addition of 99 does not. If $m > n$, then $99 - m < 99 - n$.

Solutions of quadratic equations

The solution to the equation $\frac{1}{1-x} \times \frac{4-4x}{x} = 4$ is $x = 1$. True or false?

On being presented with this problem then you would probably seek to replace the fractional components by multiplying both sides of the equation by the product $x(1-x)$. This would lead to:

$$1 \times (4 - 4x) = 4x(1 - x)$$

which, upon multiplying out the brackets, leads to:

$$4 - 4x = 4x - 4x^2$$

Upon transposition this leads to:

$$4x^2 - 8x + 4 = 0$$

and hence

$$x^2 - 2x + 1 = 0$$

This may be solved by factorising as:

$$(x - 1)^2 = 0$$

and so $x = 1$. So the answer to the original problem would appear to be 'true'.

It is always good practice to put the solution back into the original problem to confirm that it is correct. With $x = 1$ this will give:

$$\frac{1}{1-1} \times \frac{4-4\times 1}{1} = 4$$

which simplifies to $\frac{1}{0} \times \frac{0}{1} = 4$ and so to $\frac{0}{0} = 4$, which is nonsense because $\frac{0}{0}$ is indeterminate.

So what went wrong?

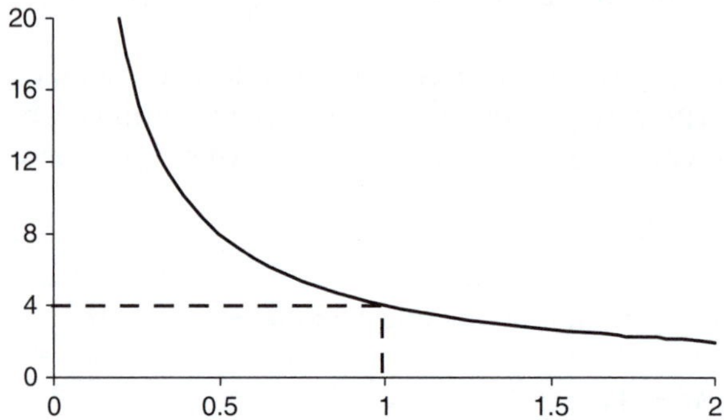

If we draw a graph of $y = \frac{1}{1-x} \times \frac{4-4x}{x}$ it looks like this and it appears to be a smooth line, even where $x = 1$ and $y = 4$.

However, the solution that $x = 1$ can still be justified using L'Hôpital's rule which says that if $y = \frac{f(x)}{g(x)} = \frac{0}{0}$ for a given value of x then, at that value, $y = \frac{f'(x)}{g'(x)}$. See the section on 'Zero and infinity'. Since:

$$y = \frac{1}{1-x} \times \frac{4-4x}{x} = \frac{4-4x}{(1-x)x} = \frac{4-4x}{x-x^2} = \frac{f(x)}{g(x)}$$

in this case, when $x = 1$:

$$y = \frac{f'(x)}{g'(x)} = \frac{-4}{1-2x} = \frac{-4}{1-2\times 1} = 4$$

Quiz

Find all solutions to the following equations and confirm your solutions by putting each solution back into the original equation.

1. $\dfrac{x-3}{x^2-9}=1$
2. $\dfrac{\sqrt{x}+1}{x-1}=2$

---

1. $x=-2$ only.

Multiply both sides of the equation by x^2-9 to get $x-3=x^2-9$.

Transpose into quadratic form, $x^2-x-6=0$.

Solve by factorising, $(x-3)(x+2)=0$.

Hence the solutions are $x=3$ and $x=-2$.

Checks. Put $x=-2$ back into the original equation; $\frac{-2-3}{(-2)^2-9}=\frac{-5}{-5}=1$ and therefore $x=-2$ is a correct solution.

Put $x=3$ back into the original equation; $\frac{3-3}{3^2-9}=\frac{0}{0}$ is undefined.

Use L'Hôpital's rule $y=\frac{f'(x)}{g'(x)}=\frac{1}{2x}=\frac{1}{6}$ which is not 1, therefore $x=3$ is not a solution to this problem.

If the original equation had been factorised as $\frac{x-3}{(x-3)(x+3)}=1$ then, upon cancelling, this becomes $\frac{1}{(x+3)}=1$. Hence, upon multiplying through, $x+3=1$ and so the only solution is $x=-2$.

2. $x=\frac{9}{4}$ only.

Multiply both sides of the equation by $x-1$ to get $\sqrt{x}+1=2x-2$.

Subtract 1 from both sides, $\sqrt{x}=2x-3$.

Square both sides, $x=4x^2-12x+9$.

Transpose into quadratic form and divide both sides by 4, $x^2-\frac{13}{4}x+\frac{9}{4}=0$.

Solve by factorising, $(x-1)\left(x-\frac{9}{4}\right)=0$.

Hence the solutions are $x=1$ and $x=\frac{9}{4}$.

Checks. Put $x=\frac{9}{4}$ back into the original equation; $\frac{\sqrt{\frac{9}{4}}+1}{\frac{9}{4}-1} = \frac{\frac{3}{2}+1}{\frac{9}{4}-1} = \frac{\frac{5}{2}}{\frac{5}{4}} = 2$ and therefore $x=\frac{9}{4}$ is a correct solution.

Put $x=1$ back into the original equation; $\frac{\sqrt{1}+1}{1-1} = \frac{\sqrt{1}+1}{0}$ which is undefined.

Therefore, $x=1$ is not a solution.

If the original equation had been factorised as $\frac{\sqrt{x}+1}{(\sqrt{x}+1)(\sqrt{x}-1)} = 2$ then, upon cancelling, this becomes $\frac{1}{(\sqrt{x}-1)} = 2$.

Hence, upon multiplying through, $1 = 2(\sqrt{x}-1)$ and hence $\frac{1}{2} = \sqrt{x}-1$, so $\frac{3}{2} = \sqrt{x}$, and $x=\frac{9}{4}$ is the only solution.

Solutions to simultaneous quadratic equations

We know that a pair of independent simultaneous equations in two variables has one solution for each of the variables. We also know that a quadratic equation can have up to two distinct real solutions. Therefore, will a pair of simultaneous quadratic equations in x and y have up to two distinct solutions for the pair (x, y)?

For example, given the pair of independent simultaneous quadratic equations $2x^2+y^2=3$ and $2y^2+x^2=3$, then the solutions for (x, y) are $(1, 1)$ and $(1, -1)$. We can verify that these are true by putting $x=1$ and $y=1$ back into both the original equations and then doing the same with $x=1$ and $y=-1$ and see that everything works. So we have solved the equations, yes?

Not really, because we have only found half the answer. With a pair of independent simultaneous quadratic equations in two variables there are up to *four* distinct real solutions.

Here is how you might find them:

Make x^2 the subject of $2y^2+x^2=3$	$x^2=3-2y^2$
Substitute for x^2 in $2x^2+y^2=3$	$2(3-2y^2)+y^2=3$
Hence	$6-4y^2+y^2=3$
Simplify to	$y^2=1$
So	$y=\pm 1$
Substitute 1 for y^2 in $2y^2+x^2=3$	$2+x^2=3$
Hence	$x^2=1$
So	$x=\pm 1$

But the solutions for x and y are not related, so either solution for x can go with either solution for y. Therefore, the four solutions for (x, y) are $(1, 1)$, $(1, -1)$, $(-1, 1)$ and $(-1, -1)$.

We can illustrate this by plotting the curves of $2x^2+y^2=3$ and $2y^2+x^2=3$. Both are equations of ellipses; one is oblate (squashed in the y-direction) and the other is prolate (squashed in the x-direction). The ellipses intersect at four points; the four solutions to the original pair of independent simultaneous quadratic equations.

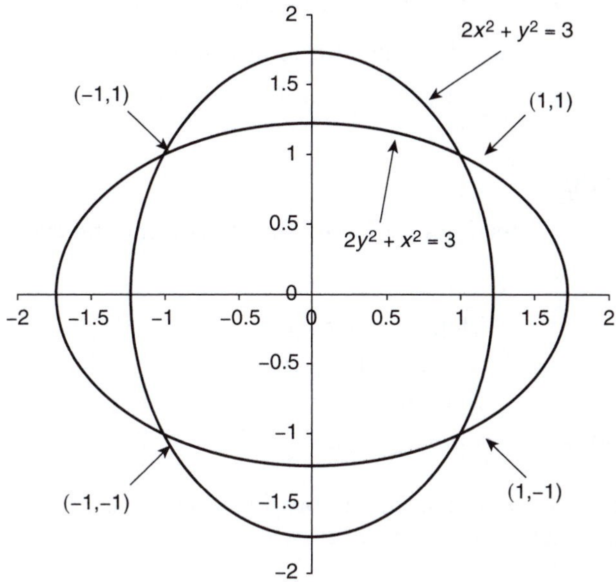

Quiz

Find all the real and distinct solutions to the following pairs of simultaneous equations. Try sketching a graph to verify your solutions.

1. $y = 2x + 3$ and $y = -x + 9$
2. $x^2 + y^2 - 4x - 6y - 12 = 0$ and $7y + x - 48 = 0$
3. $x^2 + y^2 = 8$ and $y^2 = 4x - 4$
4. $x^2 + y^2 - 28x + 96 = 0$ and $y^2 = 12x - 36$

1. $(2, 7)$. Two straight lines cross at one point.
2. $(-1, 7)$ and $(6, 6)$. A straight line crosses a circle at two points.
3. $(2, 2)$ and $(2, -2)$. A parabola intersects a circle at two points.
4. $(6, 6)$, $(6, -6)$, $\left(10, \sqrt{84}\right)$ and $\left(10, -\sqrt{84}\right)$. A parabola intersects a circle at four points.

Logarithms

Equations sometimes contain the logarithm function. For example:

If	$2\log_{10} x = \log_{10} 16$
then	$\log_{10} x^2 = \log_{10} 16$
so	$x^2 = 16$
and hence	$x = \pm 4$

Each line follows from the previous line so why should we consider that there is a mistake?

The problem may be resolved by taking the answer and putting it back into the original problem. It is true that

$$2\log_{10} 4 = \log_{10} 16$$

But no calculator will cope with $2\log_{10}(-4)$ because there is no real value for the logarithm of a negative number. Since $\log_{10}(-4)$ does not have a real value, then -4 cannot be a solution for x.

How long does it take you to spot the error in this?

A, B and C are positive variables.

$$\text{If} \quad x = \log A - \log B + \log C$$

$$\text{Then} \quad x = \log \frac{A}{BC}$$

This was just a misapplication of signs. The correct answer is of course $x = \log \frac{AC}{B}$.

Here is something a little more complicated; see if you can spot the error in this derivation, given this opening line:

$$3\log_5 x = 4\log_x 5 + 1$$

$$\begin{aligned} \log_5 x^3 &= \log_x 5^4 + \log_5 5 \\ &= \log_x 625 + \log_5 5 \\ &= \frac{\log_5 625}{\log_5 x} + \log_5 5 \\ &= \log_5 \left[\frac{625}{x} \times 5\right] \\ &= \log_5 \frac{3125}{x} \end{aligned}$$

$$\text{So} \quad \log_5 x^3 - \log_5 \frac{3125}{x} = 0$$

$$\log_5 \frac{x^4}{3125} = 0$$

$$x^4 = 3125$$
$$x \approx 7.48$$

Line 5, the one with the square brackets, is where the error lies, because $\log_a\left(\frac{b}{c}\right) \neq \frac{\log_a b}{\log_a c}$. It is $\log_a\left(\frac{b}{c}\right) = \log_a b - \log_a c$.

Before we look at a better route to the solution of the original equation we need to derive a particular relationship. The formula for changing the base of a logarithm is $\log_a b = \frac{\log_c b}{\log_c a}$. If $c = b$, then this formula becomes $\log_a b = \frac{\log_b b}{\log_b a}$ so $\log_a b = \frac{1}{\log_b a}$. You could verify this on a calculator by finding $\log_{10} e$ and $\frac{1}{\ln 10}$, which are the same. So in our original problem:

$$3\log_5 x = 4\log_x 5 + 1$$
$$3\log_5 x = \frac{4}{\log_5 x} + 1$$

Multiply through by $\log_5 x$ and gather terms on the left.

$$3\left(\log_5 x\right)^2 - \log_5 x - 4 = 0$$

This is a quadratic equation in $\log_5 x$. Solve by factorising.

$$\left(3\log_5 x - 4\right)\left(\log_5 x + 1\right) = 0$$

This suggests two solutions: $\log_5 x = -1$, which does not have a real solution, and

$$\log_5 x = \frac{4}{3}.$$

So $\quad x = 5^{\frac{4}{3}} \approx 8.55.$

Quiz

Are the following statements correct?

1. If $\log_{10} y = x + \log_{10} x$ then $y = 10^x x$.
2. $\log_{10}(x - y) = \log_{10} x - \log_{10} y$.
3. $\log(\sin x) = \sin(\log x)$.
4. $\log_{10} x \log_{10} y = \log_{10}(xy)$.
5. $\log_a(b^3) \times \log_b(c^4) \times \log_c(a^5) = 60$.

1. Yes. Hint: start by subtracting $\log_{10} x$ from both sides.
2. No, there is no simplification for this.
3. No, you cannot interchange functions; it is not like the rules of multiplying or even the power of a power rule.
4. No, there is no simplification for this.
5. Yes. Hint: start by converting all logarithms to the same base, say e; $3\dfrac{\ln b}{\ln a} \times 4\dfrac{\ln c}{\ln b} \times 5\dfrac{\ln a}{\ln c} = 60$ and all the logarithm terms cancel.

Zero and infinity

To say that zero and infinity, which is written as ∞, are the two numbers that create the most difficulty in arithmetic and algebra would not be true. And that is simply because although zero is a number, infinity is not. And even that is dubious in that it is questionable whether zero is a number or really just the absence of something that is not zero.

Infinity is not a number because, by definition, it means 'unbounded'. If you think you have thought of a number that is so large that it can be treated as infinite, then by definition you have not, because you cannot begin to imagine a finite number so large that it is even one step on the infinitely long journey towards infinity.

Working with zero and infinity often produce apparent anomalies. Here are some examples.

The reciprocal of infinity

As $x \Rightarrow \infty$ then $\frac{1}{x} \Rightarrow 0$, where the symbol $\Rightarrow$ means 'tends to'. One might be tempted to say that 1 divided by infinity is 0 and 1 divided by minus infinity is also 0. That might lead one to think that

$$\frac{1}{-\infty} = 0 = \frac{1}{\infty}$$

If this was true that would mean that $\infty = -\infty$. If real numbers are thought of as being in a continuum, from decreasing negative numbers through zero to increasing positive numbers, then continuously decreasing negative numbers are heading in the direction of $-\infty$ and increasing positive numbers are heading in the direction of $+\infty$. $+\infty$ and $-\infty$ are definitely not at the ends of the continuum, they are merely unattainable goals. So, since $+\infty$ and $-\infty$ are in opposite directions along the continuum of numbers, their inverses cannot both be the same while they are not.

One way to get around this problem is to note that as a number tends to infinity its inverse tends to zero. Therefore, depending upon whether the number is tending towards $+\infty$ or $-\infty$, its inverse will tend towards zero from the positive direction or the negative direction. In that case, we could say that

$$\text{as } x \Rightarrow -\infty \text{ then } \frac{1}{x} \Rightarrow 0^- \text{ and}$$

$$\text{as } x \Rightarrow +\infty \text{ then } \frac{1}{x} \Rightarrow 0^+$$

Although 0^- and 0^+ are both zero they are different kinds of zero in that they still maintain a positive or negative characteristic. So although they both have the same value, zero, and zero is neither positive nor negative, they retain an attribute of either positiveness or negativeness.

Product of zero and infinity

What is the product of zero and infinity? It is often erroneously claimed that since

$$\frac{1}{\infty} = 0 \text{ then, by multiplying both sides of the equation by } \infty,$$

we get

$$0 \times \infty = 1$$

Is this true? Consider the case if we started with $\frac{2}{\infty} = 0$, multiplied both sides of the equation by ∞ to get $0 \times \infty = 2$; that could be taken as proof that since $0 \times \infty = 1$ and now $0 \times \infty = 2$ then $1 = 2$. Clearly nonsense; so what went wrong?

Any number, no matter how large, when multiplied by 0, is 0. An infinite replication of any positive number, no matter how small, is infinite. So what is $0 \times \infty$? Remember that although 0 is a number, ∞ is not, so $0 \times \infty$ is in fact undefined; that is, there is no answer to this problem.

Division of zero or infinity by itself

In arithmetic, we are familiar with identifying common factors in a fraction and then simplifying that fraction by cancelling the common factor from both numerator and denominator. So can we do the same if the common factors are zero or infinity, as here?

$$\frac{\infty}{\infty} = 1 = \frac{0}{0}$$

The straight answer is no. Dividing ∞ by ∞ is meaningless because ∞ is not a number. Dividing 0 by 0 is also undefined. Try it on your calculator and it will show a message like 'Maths Error'.

However, there may be a solution to the problem of 0 divided by 0, depending upon how that expression was originally formed. For example, if the problem is to find y when $x = 0$ for the function

$$y = \frac{3x}{\sin x}$$

then, on putting $x = 0$ into the equation, you get $y = \frac{0}{0}$ which appears to be undefined.

The solution is through the use of L'Hôpital's rule, which says that if $y = \frac{f(x)}{g(x)} = \frac{0}{0}$ for a given value of x then $y = \frac{f'(x)}{g'(x)}$, where $f(x)$ and $g(x)$ are functions of x and where $f'(x)$ and $g'(x)$ are $f(x)$ and $g(x)$ differentiated with respect to x and evaluated with the same value of x. The proviso is that $y = \frac{f'(x)}{g'(x)} \neq \frac{0}{0}$. If it does, just apply L'Hôpital's rule to $y = \frac{f'(x)}{g'(x)}$ to get $y = \frac{f''(x)}{g''(x)}$, and so on.

So, applying L'Hôpital's rule to the example above when $x = 0$ leads to:

$$y = \frac{3x}{\sin x} = \frac{3}{\cos x} = \frac{3}{1} = 3$$

Infinitely meaningless

Given that ∞ is not a number then:

> $\infty + \infty \neq 2\infty$ because that would imply that infinity is finite. If you want to add ∞ to ∞ the result is undefined as ∞.
>
> $\infty - \infty \neq 0$. Again that would imply that infinity can be quantified. $\infty - \infty$ is therefore undefined.
>
> $-(1 - \infty) \neq \infty - 1$. The brackets have been removed correctly but the terms $1 - \infty$ and $\infty - 1$ are meaningless as they imply that infinity can be quantified.
>
> Multiplying ∞ by itself or raising ∞ to a positive power only produces ∞ again. So, for example, $\infty \times \infty$, ∞^5 and ∞^∞ are all undefined and infinite.

Zero to the power of zero

We know that any non-negative number to the power of 0 is 1, that is $a^0 = 1$. Therefore, if $a = 0$ then $0^0 = 1$.

We also know that 0 to the power of any non-negative number is 0, that is $0^a = 0$. Therefore, if $a = 0$ then $0^0 = 0$.

So we now have contradictory solutions for 0^0, namely 1 and 0. Which one is right? If you try 0^0 in a Casio calculator you get MATHS ERROR. If you try 0^0 in Microsoft Excel you get #NUM!; both implying that 0^0 is undefined, However, the calculator in Microsoft windows returns 1 for 0^0 as do the scientific calculators at calculateforfree.com, calculator.net, motionnet.com and many other web-based calculators. This now makes three possible solutions for 0^0; 1, 0 and undefined.

If $0^0 = 0$ then $0^{-0} = \frac{1}{0} = \infty$. But since $0 = -0$ the powers are the same, so that implies that $0 = \infty$. However, if $0^0 = 1$ then $0^{-0} = \frac{1}{1} = 1$ and there is no contradiction.

So let us consider the term 0^n. If $n > 0$ then $0^n = 0$. On the other hand, if $n < 0$ then $0^n = \frac{1}{0}$. So what is 0^n if $n = 0$? It cannot be both 0 and $\frac{1}{0}$. It is impossible to say which because we have a discontinuity in the function $y = 0^x$.

OK, so which is right? The answer is that there is no definitive answer. The definition of 0^0 can be 1, 0 or undefined. Which you accept probably depends upon which is most convenient in the context of the problem you are trying to solve. All rather unsatisfactory, but not everything is as clear cut as $1 + 1 = 2$ in maths.

Quiz

Which of the following are true?

1. $2 \times 0 = 0$
2. $0^2 = 0$
3. $(0 + 1)^0 = 0^1$
4. $\infty^{-\infty} = -\infty$

5. $\dfrac{\infty}{-\infty} = -\infty$

6. $\dfrac{(4^2-17)^{16}}{\left((-3)^2-3^2\right)} = \dfrac{1}{18}$

7. $\infty^0 - 0^\infty = 1$

1. Yes.
2. Yes.
3. No, $0^1 = 0$ but $(0+1)^0 = 1^0 = 1$.
4. No, $\infty^{-\infty}$ is undefined.
5. No, $\dfrac{\infty}{\infty}$ is undefined; the minus sign makes no difference to this.
6. No, $(4^2-17)^{16} = 1$ in the numerator but $\left((-3)^2-3^2\right) = 9-9 = 0$ in the denominator. So the fraction reduces to $\dfrac{1}{0}$ which is undefined.
7. No, both terms are undefined, so the whole sum is also undefined.

Getting the right root

Consider this statement—every number is the same as every other number. Mad as that may sound, here is the 'proof'.

Let x and y be any different real numbers and let p be the average of x and y.

Therefore	$2p = x + y$
Multiply both sides of this equation by $(x-y)$	$2p(x-y) = (x+y)(x-y)$
Multiply out the brackets	$2px - 2py = x^2 - y^2$

Add $y^2 - 2px$ to both sides of the equation	$y^2 - 2py = x^2 - 2px$
Add p^2 to both sides of the equation	$y^2 - 2py + p^2 = x^2 - 2px + p^2$
And factorise	$(y - p)^2 = (x - p)^2$
Find the square root of both sides	$y - p = x - p$
Add p to both sides	$y = x$

And since x and y can be any number then every number is the same as any other number and so all numbers are the same. Clearly this is nonsense, so what is wrong?

The 'proof' is correct in that it follows the dictum 'what you do to one side of an equation you must do to the other', except for finding the square roots in the penultimate line.

It is often stated that there are two solutions for a square root; for example, $\sqrt{4} = +2$ or -2. The converse is certainly true in that the square of $+2$ is 4 and so is the square of -2. However, the implied value of a square root is its positive value only. So really $\sqrt{4} = +2$ only.

So what went wrong in the penultimate line of the 'proof' above was that if $y - p$ is the positive root of $(y - p)^2$ then $x - p$ is the negative root of $(x - p)^2$. We can see that is true by substituting $\frac{1}{2}(x + y)$ for p. In that case, the 'Find the square root of both sides' line reads as:

Find the square root of both sides	$y - \frac{1}{2}(x + y) = x - \frac{1}{2}(x + y)$
Simplify	$\frac{1}{2}(y - x) = \frac{1}{2}(x - y)$

So one side is the negative value of the other and this can only be true if $x = y$. If we return to the factorisation line where

$$(y - p)^2 = (x - p)^2$$

and substitute $\frac{1}{2}(x + y)$ for p, then:

$$\left(y - \tfrac{1}{2}(x + y)\right)^2 = \left(x - \tfrac{1}{2}(x + y)\right)^2$$

and simplify $\left(\frac{1}{2}y-\frac{1}{2}x\right)^2=\left(\frac{1}{2}x-\frac{1}{2}y\right)^2$

which is true, and the positive roots are $\frac{1}{2}(y-x)=\frac{1}{2}(y-x)$

provided that $y \geq x$. However, this just leads to 0 = 0, which although true is not very helpful in proving our statement, which in fact now cannot be proved. So it is not true that 'every number is the same as every other number', which comes as a relief, otherwise all maths would be nonsense.

Quiz

In each of the following derivations which line contains the error?

1.

	$a=0$	(1)
Multiply by $(a-1)$	$(a-1)\times a=(a-1)\times 0$	(2)
	$=0$	(3)
Divide by a	$a-1=0$	(4)
Subtract 1	$a=1$	(5)

2.

	$x=\frac{3\pi}{2}$	(1)
	$\cos x=0$	(2)
	$x=\cos^{-1}0$	(3)
	$x=\frac{\pi}{2}$	(4)
Therefore	$\frac{3\pi}{2}=\frac{\pi}{2}$	(5)
So	$3=1$	(6)

1. Error in line (4)—division is by 0
2. Error in line (4)—there are an unbounded number of solutions for x; $x=(2n+1)\frac{\pi}{2}$, where n is any integer.

Summary of the main points

- In a lengthy algebraic equation it is often convenient to simplify by replacing a sum of expressions with a product of other expressions. To do this you first need to identify common factors. Having created an expression which is a product of factors, a good check is to multiply out the product to ensure that the outcome is the original sum of expressions. For example, given $2xy + 3xz$ this may be expressed as the product $x(2y + 3z)$. Check this by multiplying the factors x and $(2y + 3z)$ to get back to $2xy + 3xz$.
- The product of two numbers which have the same base, but where each is raised to a power, is the same base raised to the sum of the powers. $x^y x^z = x^{y+z}$. But note that $x^y + x^z \neq x^{y+z}$.
- The basic logarithm relationships are:

$$\log_a(bc) = \log_a(b) + \log_a(c) \qquad \log_a(b^c) = c\log_a(b)$$

$$\log_d(x) = \frac{\log_a(x)}{\log_a(d)} \qquad \log_d(a) = \frac{1}{\log_a(d)}.$$

- When solving equations that involve logarithms remember to check that any potential solutions lie within the range of the logarithm function. That is, $\log x$ is real only if $x > 0$.
- The basic square root relationships are:

$$\left(\sqrt{x+y}\right)^2 = x + y \qquad \left(\sqrt{x} + \sqrt{y}\right)^2 = x + 2\sqrt{xy} + y$$

$$\text{but } \sqrt{x+y} \neq \sqrt{x} + \sqrt{y}$$

- For the square root of the product xy, $\sqrt{x \times y} = \sqrt{x} \times \sqrt{y}$ is true for all cases, ***except*** where both x and y have negative values.
- There is no real value for the square root of a negative number. The square root of a negative number can be expressed in complex form as the product of the square root of a positive number and i, the square root of -1. For example, $\sqrt{-7} = \sqrt{7}\,i$, where $i = \sqrt{-1}$.

- For powers of products all parts of the product must be raised to the power if the brackets are to be expanded. A similar relationship applies to roots. For example:

$$(ax)^b = a^b x^b \text{ and } \sqrt[b]{ax} = \sqrt[b]{a}\,\sqrt[b]{x}$$

- But beware of negative values of a and x, as mentioned above.
- When expanding brackets you will need to multiply all of the terms in the first bracket by all of the terms in the second bracket. For example:

$$\begin{aligned}&(a+b)(c+d) = ac+ad+bc+bd\\&(x+y)^2 = x^2+2xy+y^2\\&(x-y)^2 = x^2-2xy+y^2\\&(x+y)(x-y) = x^2-y^2\\&(a+b+c)(p+q+r) = ap+aq+ar+bp+bq+br+cp+\\&\quad cq+cr\end{aligned}$$

- If you use one of the methods of 'product by areas', 'smiley face', 'eagle and tortoise' or 'FOIL' it may help to ensure that no terms are missed or duplicated.
- The sign of a negative value raised to an integer power may be easily found by noting whether the integer power is odd or even. If even the result is positive and if odd the result is negative. A similar relationship applies to roots. For example, $(+2)^5 = 32$ and $(-2)^5 = -32$. Also, $\sqrt[3]{+27} = 3$ and $\sqrt[3]{-27} = -3$.
- Unless otherwise stated or implied in the context of a problem, the square root of a number implies only the positive root. In finding the root of a quadratic equation there are always two solutions. Both may not be distinct or both may not be real. Which ones, if any, are valid will depend on the context of the problem.
- If you have a fraction which you wish to split into a sum of fractions then both parts of the numerator must be divided by the denominator. For example, $\frac{p^3+p^2r}{p} = p^2+pr$ and $\frac{12+4.5}{3} = 4+1.5$. Be careful where the numerator and/or denominator are themselves fractions. Where the denominator is a sum of

terms it is unlikely that anything can be done to simplify the expression. For example, $\frac{pr}{p+r}$ does not simplify.

- Before undertaking the solution of a polynomial equation see if you can identify the maximum number of possible solutions. Then see if you can find them. So if the equation is a quadratic there could be two distinct real solutions, if it is a cubic there could be up to three distinct real solutions and if the highest power of the variable is 6 there could be up to six distinct real solutions.
- If an equation may be expressed as $f(x) = 0$ and $f(x)$ may be factorised, then any of the factors may be equal to 0, and hence values of x may be found. So if $x^3 + 6x^2 + 11x + 6 = 0$ this may be factorised to $(x + 1)(x + 2)(x + 3) = 0$. Therefore, $x + 1 = 0$ or $x + 2 = 0$ or $x + 3 = 0$. Beware the temptation to say that $x = 1$ or $x = 2$ or $x = 3$ because the correct solutions are $x = -1$ or $x = -2$ or $x = -3$.
- When working with functions note how the function acts upon the input to obtain the output. When working with functions of functions, generally $f(g(a)) \neq g(f(a))$.
- When 'multiplying through' an equation by a term, remember to multiply the ***whole of both sides*** of the equation by the term.
- An inequality shows when two expressions are not the same. Be careful when dividing or multiplying both sides by negative values as the inequality will change. Taking the reciprocal of both sides will have a similar effect. For example, multiplying both sides of the inequality $5 > 3$ by -2 leads to the inequality $-10 < -6$. Similarly, $5 > 3$ but $\frac{1}{5} < \frac{1}{3}$.
- When you solve equations make sure that continuity conditions remain valid.
- The solution to a pair of simultaneous quadratic equations may have up to four distinct pairs of solutions for the two variables.
- The concepts of zero and infinity must be treated with caution. As $x \Rightarrow -\infty$ then $\frac{1}{x} \Rightarrow 0^-$ and as $x \Rightarrow +\infty$ then $\frac{1}{x} \Rightarrow 0^+$.
- Infinity is unbounded. Products or sums involving infinity are generally undefined. Division by 0 is also undefined.

- If, for a particular value of x, both $f(x)$ and $g(x)$ are 0, then it would appear that $\frac{f(x)}{g(x)} = \frac{0}{0}$, which is undefined. However, a solution for $\frac{f(x)}{g(x)}$ may be found by using L'Hôpital's rule, which says that if $y = \frac{f(x)}{g(x)} = \frac{0}{0}$ for a given value of x then $y = \frac{f'(x)}{g'(x)}$. If this is still $\frac{0}{0}$ then $y = \frac{f''(x)}{g''(x)}$. If this is still $\frac{0}{0}$ then $y = \frac{f'''(x)}{g'''(x)}$, and so on.
- The definition of 0^0 can be 1, 0 or undefined; it depends upon which is most convenient in the context of the problem you are trying to solve.

9 Errors in trigonometry

Calculator's angle mode

Most scientific calculators can handle angles in degrees, radians and grad format. The different formats reflect how the circle is divided. One complete circle is 360° (degrees), 400^g (grads), 2π (radians) or 6400 mils, although that last format is seldom supported.

There are times when one angular format is more useful than another. Any calculation involving calculus, that is with differentiation or integration, invariably implies the radian format. Calculations involving trigonometry or geometry are usually best handled in any format other than radians. If the angle format is in degrees, those degrees may be in decimals of a degree, or the degrees may be subdivided into minutes and seconds. Grads may be preferred if you are working in a country or profession where this format is common, for example, as a surveyor in most of Europe. Mils are often used by the military.

The display on most calculators will show which angle format you are in. It is worth checking you are in the right format before you start a new calculation, especially if you regularly have to change format, for example, if you are a science or engineering student. To illustrate the problem:

in degrees mode	$\sin(10) = 0.1736$
in grad mode	$\sin(10) = 0.1564$
in radians mode	$\sin(10) = -0.5440$

and

in degrees mode	$\cos^{-1}(0.8) = 36°52'12''$
in grad mode	$\cos^{-1}(0.8) = 40.9666$
in radians mode	$\cos^{-1}(0.8) = 0.6435$

Quiz

1. Use your calculator to find the tangents of 60 degrees, 60 grads and 60 radians, giving your answer to 4 decimal places.
2. Use your calculator to find the arcsine ($\sin^{-1}$) of 0.65 in radian, grad and degree mode. Give your answer to 4 decimal places of a radian and grad, and to the nearest arc-second, respectively.
3. Identify by trial and error which calculator mode is in use in each of the following:

   a. $\sin(21.274) = 0.3280$
   b. $\cos(4.321) = 0.9972$
   c. $\tan(15.231) = -0.5168$
   d. $\sin^{-1}(0.123) = 7.0653$
   e. $\cos^{-1}(0.234) = 1.3346$
   f. $\tan^{-1}(0.345) = 21.1494$

1. 1.7321, 1.3764, 0.3200
2. 0.7076, 45.0462, 40° 32′ 30″
3. a. Grad b. Degree c. Radian d. Degree e. Radian f. Grad

Reciprocal of functions

Which of the following are correct?

$$\sec x = \frac{1}{\sin x} \quad \csc x = \frac{1}{\cos x} \quad \cot x = \frac{1}{\tan x}$$

In a right-angled triangle an angle, other than the right-angle, may be expressed as the ratio of any two sides of the triangle. The ratios most often used are those of the sine, cosine and tangent. They are usually abbreviated to sin, cos and tan.

In a right-angled triangle, the hypotenuse is the side facing the right-angle. The opposite and the adjacent sides are with respect to the subject angle, x in the diagram. Therefore:

$$\sin x = \frac{\text{opposite}}{\text{hypotenuse}}$$

$$\cos x = \frac{\text{adjacent}}{\text{hypotenuse}}$$

$$\tan x = \frac{\text{opposite}}{\text{adjacent}}$$

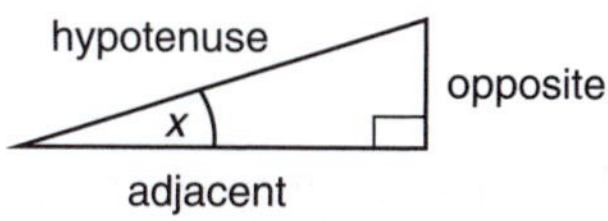

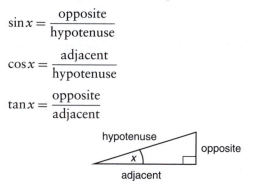

A popular mnemonic for remembering these relationships is SOH CAH TOA, which stands for:

> <u>S</u>ine is <u>O</u>pposite over <u>H</u>ypotenuse, <u>C</u>osine is <u>A</u>djacent over <u>H</u>ypotenuse, <u>T</u>angent is <u>O</u>pposite over <u>A</u>djacent.

But the ratios of sides that give us sine, cosine and tangent are not the only possible ratios. There are three others: secant, cosecant and cotangent. These are often abbreviated to sec, cosec or csc, and cot, respectively. They are the ratios:

$$\sec x = \frac{\text{hypotenuse}}{\text{adjacent}} \quad \csc x = \frac{\text{hypotenuse}}{\text{opposite}} \quad \cot x = \frac{\text{adjacent}}{\text{opposite}}$$

Relating these back to sine, cosine and tangent, it can be seen that:

$$\sec x = \frac{1}{\cos x} \quad \csc x = \frac{1}{\sin x} \quad \cot x = \frac{1}{\tan x}$$

Although most people remember that $\cot x = \dfrac{1}{\tan x}$, many confuse the other two relationships because it is tempting to think that $\csc x = \dfrac{1}{\cos x}$ since both parts begin with a 'co', and that $\sec x = \dfrac{1}{\sin x}$ since both parts begin with the letter 's'. Both of course are wrong; you need to remember them the other way around.

Quiz

Without using a calculator, can you fill in the missing (?) trigonometric functions?

1. $?(30°) = 0.5$
2. $?(60°) = 0.5$
3. $?(45°) = 1$
4. $?(60°) = 2$
5. $?(30°) = 2$
6. What are all the possible answers to $?(x) = 1$, where ? is a trigonometric function and x is an angle in the range $0° \leq x \leq 90°$?
7. What are all the possible answers to $?(x) = 0$, where ? is a trigonometric function and x is an angle in the range $0° \leq x \leq 90°$?

1. sine
2. cosine
3. tangent or cotangent
4. secant
5. cosecant
6. $\sin(90°)$, $\cos(0°)$, $\tan(45°)$, $\sec(0°)$, $\csc(90°)$, $\cot(45°)$
7. $\sin(0°)$, $\cos(90°)$, $\tan(0°)$, $\cot(90°)$

Inverse functions

The function $\cos^{-1}$ is the inverse function of cosine. So if $y = \cos x$ then $x = \cos^{-1} y$. That is that x is the angle such that its cosine is y. Similarly $\sin^{-1}$ is the inverse function of sine and $\tan^{-1}$ is the inverse function of tangent.

What is wrong with this statement $\cos^{-1}(0.5) = 60$?

This may appear to be a small point, but the omission is one that could cause confusion in later computations and would be likely to lose marks in an examination. What is missing from the right-hand side of the equals sign, after 60, is the unit: degrees, radians, mils or grads? Without units the answer is potentially meaningless.

Spot the error here:

$$\sin^{-1}(0.5) = 30°,$$

$$\sin^{-1}(-0.5) = -30°,$$

$$\cos^{-1}(0.5) = 60° \text{ and}$$

$$\cos^{-1}(-0.5) = -60°.$$

If it took a moment or two longer than you would have expected that is probably because you were momentarily tricked into accepting that there is a consistency associated with minus signs. Whereas it is true that $\sin^{-1}(-0.5) = -30°$, $\cos^{-1}(-0.5) = \pm 120°$. Try it on your calculator.

Quiz

Which of the following, rounded to 4 decimal places where appropriate, are true?

1. $\sin(36) = 0.5878$
2. $\sec^{-1}(36) = 1.5430$
3. $\tan(45) = 1.6198$
4. $\csc(-30°) = -2$
5. $\cot^{-1}(-1) = 45°$
6. $\sin\left(\csc^{-1}(-2)\right) = -2$

1. Only true if the angle units are degrees.
2. Only true if the angle units are radians. Although your calculator has the functions $\sin^{-1}$, $\cos^{-1}$ and $\tan^{-1}$, it does not have $\csc^{-1}$, $\sec^{-1}$ and $\cot^{-1}$. So to find $\sec^{-1}(36)$ that will be the same as $\cos^{-1}\left(\frac{1}{36}\right)$. Also, $\csc^{-1}x = \sin^{-1}\left(\frac{1}{x}\right)$ and $\cot^{-1}x = \tan^{-1}\left(\frac{1}{x}\right)$.
3. True; without angle units being declared, the default is normally radians because they are unitless.
4. True.
5. Not true. Correct value is $-45°$.
6. Not true. Correct value is -0.5, irrespective of the units.

Trigonometric functions

The functional nature of trigonometric functions

Here are two similar mistakes that are often made:

$$\sin(x+y) = \sin x + \sin y \text{ and } \sin 2x = 2\sin x.$$

Both indicate the assumption that sine, and by implication other trigonometric functions, may be treated as variables themselves rather than as functions of variables. In the early learning of trigonometry the functional nature of these trigonometric quantities is often misunderstood.

The correct expressions for the above trigonometric identities are:

$$\sin(x+y) = \sin x \cos y + \sin y \cos x \text{ and } \sin 2x = 2\sin x \cos x.$$

Another function where this misconception is sometimes made is with the logarithm function where it is erroneously assumed that $\log(y) = \log \times y$; the expression on the right-hand side of the equals sign is meaningless.

Power of a trigonometric function

$\sin^2 x = \sin x^2$ true or false?

Ambiguous would be the best answer. $\sin x^2$ could be interpreted in two ways: it could be the square of $\sin x$, that is $(\sin x)^2$, or it could be the sine of x^2, that is $\sin\left(x^2\right)$. So if you want the square of the sine of x write it as $\sin^2 x$, but if you want the sine of x^2 write it as $\sin\left(x^2\right)$.

Inverse function or reciprocal?

The reciprocal of the power of a trigonometric function is written, for example, as $\sin^{-3} x$ which means the reciprocal of the cube of the sine of x, that is $\dfrac{1}{\sin^3 x}$.

$\cos^{-2} x$ means the reciprocal of the square of the cosine of x, that is $\dfrac{1}{\cos^2 x}$. So the pattern suggests that $\tan^{-1} x$ would be the reciprocal of the tangent of x, that is $\dfrac{1}{\tan x}$.

This is not so. The index of -1 associated with a trigonometric function has a special meaning. It is the inverse of the function concerned, not the reciprocal. For example, $y = \tan^{-1} x$ means the value of the angle y such that its tangent is x. If $y = \tan^{-1} x$ then $x = \tan y$.

How then would you express the reciprocal of $\tan x$? Since $\tan^{-1} x$ has a special meaning then the reciprocal of $\tan x$ would have to be $(\tan x)^{-1}$. The same applies to $\sin^{-1} x$ and $\cos^{-1} x$ and the other trigonometric functions.

Just for fun

And finally, this piece of pseudo-mathematical whimsy needs no explanation, hopefully:

$$\frac{\sin x}{n} = \frac{si\not{n}x}{\not{n}} = six = 6$$

Quiz

Which of the following are correct?

1. $\tan^3 x = (\tan x)^3$
2. $\tan 2y = 2\tan y$
3. $\tan^{-1} z = (\tan z)^{-1}$
4. $\cos 2x = \cos^2 x - \sin^2 x$
5. $\cos t = \cot s$

1. Yes.
2. No, $\tan 2y = \dfrac{2\tan y}{1 - \tan^2 y}$.
3. No, if $x = \tan^{-1} z$ then $z = \tan x$, $(\tan z)^{-1} = \dfrac{1}{\tan z}$.
4. Yes.
5. No, this is nonsense.

Missing solutions

All trigonometric functions of x have two solutions in the range $0° \leq x < 360°$. For example, $\sin^{-1}(0.5)$ can be 30° or 150°, $\cos^{-1}(0.5)$ can be 60° or 300° and $\tan^{-1}(1.0)$ can be 45° or 225°. Remember that your calculator will only give you one solution and you may need to find the other by knowing the form of the trigonometric function concerned.

To find the alternative solutions, follow these rules. If:

$y = \sin^{-1} x$ the other solution may be found as $180° - y$
$y = \cos^{-1} x$ the other solution may be found as $360° - y$
$y = \tan^{-1} x$ the other solution may be found as $180° + y$

If any of these give a solution outside the range $0° \le y < 360°$ just add or subtract a further 360° as appropriate. For example:

$$\sin^{-1}(-0.1234) = -7.0884°.$$

The alternative solution is:

$$180° - (-7.0884°) = 187.0884°.$$

As $-7.0884°$, the first solution, is not within range add 360° to get 352.9116°.

If $\sin^2 x = 0.25$ and $0° \le x < 360°$ then the solutions for x are 30° and 150°. Is that correct? What, if anything, is at fault with the answer?

These two values are indeed solutions for x but they are not the only solutions. The first step to find the solution to $\sin^2 x = 0.25$ would be to find the square root of both sides of the equation, then to find $\sin^{-1}$ of both sides to evaluate x. There are of course two values for $\sqrt{0.25}$ so do not forget the non-principal solution, the negative value -0.5 as well as 0.5. This leads to two further solutions for x, which are 210° and 330°.

Quiz

How many real solutions for x are there to the following expressions where a is specified but $-1 \le a \le 1$ and $a \ne 0$?

1. $a = \cos x$, where $0° \le x < 360°$
2. $a = \tan^2 x$, where $0° \le x < 360°$
3. $a = \sin^3 x$, where $0° \le x < 360°$
4. $a = \sin x$, where $0° \le x < 720°$

5. $a = \cos^2 x$, where $0^\circ \le x < 720^\circ$
6. $a = \tan^3 x$, where $33^\circ \le x < 573^\circ$
7. $a = \cos^3 x$, where $33^\circ \le x < 573^\circ$

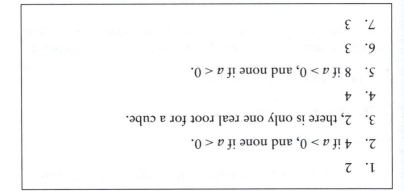

1. 2
2. 4 if $a > 0$, and none if $a < 0$.
3. 2, there is only one real root for a cube.
4. 4
5. 8 if $a > 0$, and none if $a < 0$.
6. 3
7. 3

Impossible triangles

In an examination, the phrase 'solve the triangle' means find all the angles and side lengths other than those that are given. Can you solve the triangle in the diagram where $B = 53.0^\circ$ and $c = 15.20$ m?

Here is a solution:

$$\tan B = \frac{b}{c} \text{ so } b = c \tan B = 20.17\,\text{m}$$

$$\cos B = \frac{c}{a} \text{ so } a = \frac{c}{\cos B} = 25.26\,\text{m}$$

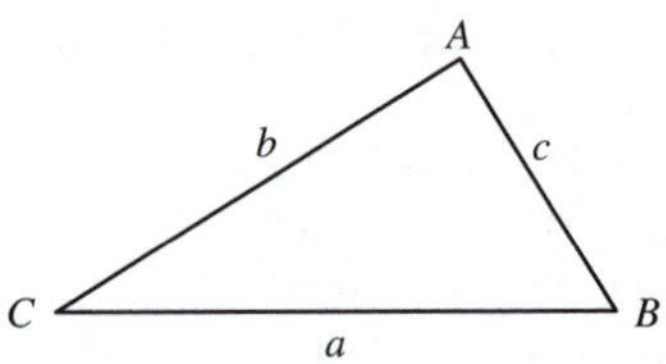

and use the cosine rule to find C

$$C = \cos^{-1}\left(\frac{a^2 + b^2 - c^2}{2ab}\right) = 37.0°$$

Did you spot the mistake?

The fact that an angle looks like a right-angle does not mean that it is one, if that fact is not explicitly stated. So how can you solve the triangle in the diagram? The answer is that you cannot; there is insufficient information. To solve any triangle you need three parts (angles or sides) including at least one side. In the problem above you only have two parts, angle B and side c, so no unique solution is possible.

Quiz

In the section above an impossible triangle was presented. Can you say why the following plane triangles, with angles A, B and C and the sides opposite the angles respectively a, b and c, are also impossible?

1. $A = 123.0°$, $a = 5.34$ m and $C = 59.0°$.
2. $A = 92.0°$, $a = 16.5$ m and $b = 18.9$ m.
3. $A = 47.0°$, $B = 56.9°$, $a = 4.23$ m and $b = 5.78$ m.

Can you solve the following triangles which all have angles A, B and C and opposite sides a, b and c?

4. $A = 25.2°$, $a = 3.45$ m and $b = c$.
5. $A = 46.7°$, $A \neq B \neq C$, $a \neq b \neq c$ and $c = 42.18$ m.
6. $a = b$, $c = 3.67$ m and $B = C$.

1. The two given angles already add up to more than 180°, so at least one of them must be in error.

2. In any triangle the smallest angle is opposite the shortest side, the middle angle is opposite the middle side and the largest angle is opposite the longest side. Since side b is longer than side a and angle A is 92°, then angle B must be greater than 92°. Therefore, the sum of the angles on the triangle will exceed 180°.
3. This information is incompatible. Two angles with their opposite sides are given. By the sine rule $b = a\dfrac{\sin B}{\sin A}$. Given the values for a, A and B, that makes $b = 4.85$ m. The information given must contain at least one error.
4. Yes. The fact that the sides $b = c$ means the triangle is isosceles and therefore angles $B = C$. $B + C = 180 - A$ so $B = C = 77.4°$. By the sine rule $b = c = 3.45\dfrac{\sin 77.4°}{\sin 25.2°} = 7.91$ m.
5. No, the relationships with the not equal, $\neq$, signs add no useful information, therefore there is insufficient data given to solve the triangle.
6. Yes. The fact that the sides $a = b$ means that angles $A = B$. Also, since $B = C$, all the angles are the same. Therefore, the triangle is equilateral, so $A = B = C = 60°$ and $a = b = c = 3.67$ m.

The sine rule

In solving triangle ABC, where $A = 23.4°$, $a = 6.58$ m and $b = 7.12$ m the following computation was carried out.

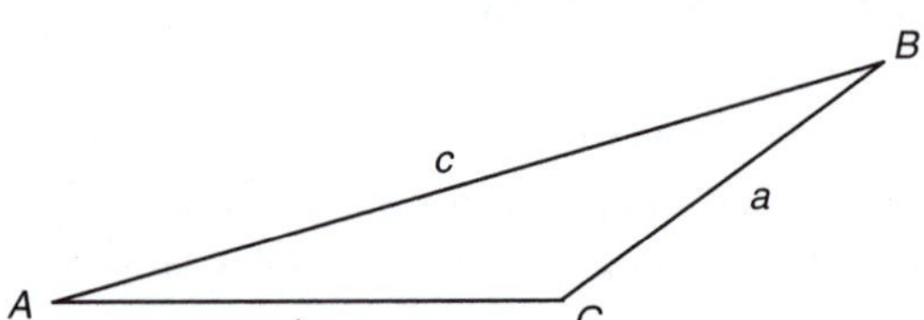

By the sine rule:

$$\sin B = \frac{b}{a}\sin A = 0.4297 \text{ hence } B = 25.45°$$

$$C = 180° - A - B = 131.15°$$

By the sine rule:

$$c = a\frac{\sin C}{\sin A} = 12.48\,\text{m}$$

The answers and the given data agree with the diagram and show the calculations are correct, so what is amiss with this solution?

The limitation with this answer is that it is not the only correct one. There is another but completely different triangle that also fits the given data. In the first line of the solution the value of $\sin^{-1}(0.4297)$ was taken to be 25.45° because that is what the calculator gives. Any angle in a triangle must be less than 180° but 25.45° is not the only solution to $\sin^{-1}(0.4297)$ that is less than 180°. The other is 180° – 25.45° = 154.55°. So the alternative solution for the triangle is:

By the sine rule:

$$\sin B = \frac{b}{a}\sin A = 0.4297 \text{ hence } B = 154.55°$$

$$C = 180° - A - B = 2.05°$$

By the sine rule:

$$c = a\frac{\sin C}{\sin A} = 0.59\,\text{m}$$

The ambiguity in the solution of the triangle arose because there are always two solutions possible when using the sine rule. Does

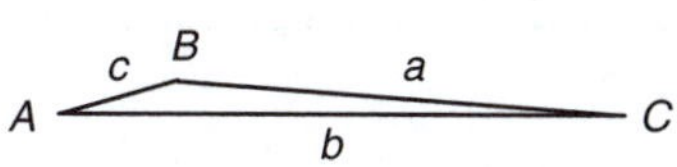

the same problem occur when using the cosine rule? For an angle which is less than 180°, its cosine lies between +1 and −1 and for any cosine there is only one possible value for that angle in the range 0°–180°. So the answer is no.

Quiz

Can you solve these triangles?

1. ABC, where $A = 35.6°$, $a = 7.82$ m and $b = 9.62$ m.
2. ABC, where $A = 28.36°$, $a = 17.62$ m and $b = 15.62$ m.
3. ABC, where $A = 42.5°$, $a = 13.61$ m and $b = 28.71$ m.

1. Yes.

$$\sin B = \frac{b}{a}\sin A = 0.7161$$

hence $B = 45.73°$ or $B = 134.27°$

$C = 180° - A - B$ $\quad$ $C = 98.67°$ or $C = 10.13°$

$c = a\dfrac{\sin C}{\sin A}$ $\quad$ $c = 13.28$ m or $c = 2.36$ m

2. Yes, but with a limitation.

$$\sin B = \frac{b}{a}\sin A = 0.4211$$

hence $B = 24.90°$ or $B = 155.10°$

$C = 180° - A - B$ $\quad$ $C = 126.74°$ or $C = -3.46°$

The second value for C is impossible for an angle in a triangle.

$$c = a\frac{\sin C}{\sin A} \qquad c = 29.73\text{ m}$$

There is only one possible solution for this triangle.

3. No. $\sin B = \dfrac{b}{a}\sin A = 1.4251$. This is out of range for the sine of an angle so the triangle cannot exist. If you try to draw it with ruler and a protractor you will find the task impossible.

Summary of the main points

- Make sure your calculator is in the correct angular mode, degrees, radians or grads, before you start any calculations.
- The reciprocals of the three common trigonometric functions are:

$$\sec x = \frac{1}{\cos x} \quad \csc x = \frac{1}{\sin x} \quad \cot x = \frac{1}{\tan x}$$

- Do not be confused by the 'co' in cosine and cosecant. Also:

$$\sec^{-1} x = \cos^{-1}\left(\frac{1}{x}\right) \quad \csc^{-1} x = \sin^{-1}\left(\frac{1}{x}\right)$$

$$\cot^{-1} x = \tan^{-1}\left(\frac{1}{x}\right)$$

- Make sure you include the units in your solutions of inverse trigonometric functions. Angles will have units of degrees, radians or grads.
- An index associated with a trigonometrical function can have specific or ambiguous meaning.
- $\sin x^2$ is ambiguous; it could be the square of $\sin x$, that is, $(\sin x)^2$, or it could be the sine of x^2, that is, $\sin\left(x^2\right)$.
- The square of the sine of x is usually written as $\sin^2 x$.

- The reciprocal of $\sin^3 x$ is written as $\sin^{-3} x$. It means $\dfrac{1}{\sin^3 x}$. Likewise, $\sin^{-2} x$ means $\dfrac{1}{\sin^2 x}$. However, $\sin^{-1} x$ is the inverse function. $y = \sin^{-1} x$ means the value of the angle y such that its sine is x. The reciprocal of $\sin x$ would be written as $(\sin x)^{-1}$.
- There are two solutions for $\sin x$, $\cos x$ and $\tan x$ in the range $0 \leq x < 360°$. $\sin^2 x$, etc. will have four solutions in the same range.
- You need three parts (angles or sides), including at least one side, to solve any triangle. Even if you have two sides and an angle there may be 2, 1 or no solutions to the triangle. It is a good idea to start by drawing a correctly dimensioned diagram to see which is most likely.

10 Calculator errors

Calculators make it feasible to do computations so much more quickly than was possible before they were invented. They also allow levels of accuracy that would be extremely tedious to achieve by other means. Talk to your older colleagues or lecturers about using slide rules and log tables for complicated or precise computations. But, however good they are, calculators have their limitations and you should be aware of them.

Factorials

A factorial of an integer is simply the product of all the positive integers up to and including the integer concerned. So, for example, $5! = 1 \times 2 \times 3 \times 4 \times 5 = 120$.

Find the factorial button on your calculator, which probably looks like $x!$, and see how large a number you can find the factorial for. If your calculator has a maximum exponent of 99 then the greatest factorial is that of 69; $69! \approx 1.711 \times 10^{98}$ (but $70! \approx 1.198 \times 10^{100}$). If your calculator has a maximum exponent of 999 then the greatest factorial possible is that of 449; $449! \approx 3.852 \times 10^{997}$ (but $450! \approx 1.733 \times 10^{1000}$).

Range

Use your calculator, in degrees mode, to find the following and you should get the values shown.

sine of 125°	= 0.81915
cosine of 305°	= 0.57358
tangent of 235°	= 1.42815

Now try the reverse process to find

arcsine of 0.81915	= 55°
arccosine of 0.57358	= 55°
arctangent of 1.42815	= 55°

In each case you will get the answer of 55°, which is not any of the angles you started with. Is your calculator wrong? No, but you have discovered one of its limitations.

The inverse trigonometrical functions of sine, cosine and tangent each have two solutions for x in the range $0 \leq x < 2\pi$ radians (or $0° \leq x < 360°$) but your calculator returns only one of them. Your calculator manual will state the range concerned; possibly $-\frac{\pi}{2} \leq x < \frac{\pi}{2}$ (or $-90° \leq x < 90°$) for sine and tangent and $0 \leq x < \pi$ (or $0° \leq x < 180°$) for cosine. So beware, there could be a 50% chance that your calculator is returning the wrong answer for your particular problem and you will need to check your answer against the original problem to confirm that it is at least realistic.

If the answer you get is clearly wrong, then:

If it was from an arcsine, subtract the answer from π (or 180°).
If it was from an arccosine, subtract the answer from 2π (or 360°).
If it was from an arctangent, add π (or 180°) to the answer.

Precision 1

The sine of any multiple of π is 0. $\sin 0 = 0$, $\sin \pi = 0$ and $\sin(n\pi) = 0$, where n is an integer. However, something rather strange may occur where n is large. The following are from the Casio fx-85MS calculator, in radians mode:

$$\sin \pi = 0 \qquad \sin 2\pi = 0$$

$$\sin 9999\pi = 5 \times 10^{-8} \qquad \sin 10000\pi = 0$$

$$\sin 99999999\pi = 4 \times 10^{-4} \qquad \sin 100000000\pi = 0$$

Precision 2

If you apply a function to a number and then apply the function's inverse to the result you should get back to the number you started with. So, for example, if you square 2 you get 4 and if you then find its square root you get back to 2. If you find the tangent of $1/4\pi$ you get 1; now find the arctangent of that result and you will get back to $1/4\pi$. If functions and inverse functions are nested, the same applies. So the arcsine of the fourth root of the fourth power of the sine of 1.23 radians will be 1.23.

$$\sin^{-1}\left(\sqrt[4]{[\sin(1.23)]^4}\right) = 1.23$$

This could appear in a calculator as:

sin^{-1}(((sin(1.23))^4)^0.25)

In theory there is no limit to the depth of nesting that is possible. So, with the calculator in degrees mode, if you

start with the number 1,
divide by 88.8,
find the sine of the result,
find its cosine,
square it,
find the square root,
find its arccosine,
find its arcsine,
and multiply by 88.8.

You should get back to the number 1 that you started with.

Try it on your calculator and see. On the Casio fx-85MS calculator the result is 1.0098. This gives a result that is about 1% in error. What went wrong?

In the midst of the computation, the cosine is of a very small number and so the cosine has a value very close to 1. The square and square root do not add to the problem significantly but the arccosine of a number close to 1 is highly unstable. Hence, rounding errors within the calculator become amplified in this particular computation.

Precision 3

Your calculator will not always give exact values. It will, for example, show that in degrees mode $\sin(30) = 0.5$ exactly, but what it shows for $\sin(60)$ will depend upon the calculator settings. If the display is fixed to four decimal places then it will return $\sin(60) = 0.8660$ and the temptation will be to believe that to be correct. However, if it is set to nine decimal places then it will return $\sin(60) = 0.866025404$. Again, this is not correct, only rounded to nine decimal places. The true value is $\sin(60°) = \frac{\sqrt{3}}{2}$, but this is an irrational number and so there is no number of decimal places that will give the true answer. For what it is worth, $\sin(60°)$ to 15 decimal places is 0.866025403784439.

Again, π to nine places in a calculator would be 3.141592654 but it may well be held to a few more places. You can test the precision of π on your calculator by setting the display to 'Scientific' with the maximum number of places available and then entering $\pi - 3.14159$ in it and see how many further digits appear.

However, π is known to many millions of places. Here, for interest, are the first one thousand. $\pi =$

3.141592653589793238462643383279502884197169399375105820974944592307
81640628620899862803482534211706798214808651328230664709384460955058223172
53594081284811174502841027019385211055596446229489549303819644288109756659334461284
756482337867831652712019091456485669234603486104543266482133936072602491412737245870066063155
8817488152092096282925409171536436789259036001133053054882046652138414695194151160943305727036575959195309
2186117381932611793105118548074462379962749567351885752724891227938183011949129833673362440656643086021394946395224737190702
17986094370277053921717629317675238467481846766940513200056812714526356082778577134275778960917363717872146844090122495343014654958537105079227968925
89235420199561121290219608640344181598136297747713099605187072113499999983729780499510597317328160963185950244594553469083026425223082533446850352619
311881710100031378387528865875332083814206171776691473035982534904287554687311595628638823537875937519577818577805321712268066130019278766111959092164201989

There are several sites on the web where π is stated to much greater precision.

Quiz

Use your calculator to make the following nested calculations. Can you say why you do not come back to the number that you started with?

1. Set your calculator to radians mode.

   Start with 3
   Divide by 2
   Find the tangent of the result
   Find the cosine of the result
   Find the arccosine of the result
   Find the arctangent of the result
   Multiply by 2

   The result should be 3 exactly.
2. Set your calculator to radians mode.

   Start with 11
   Subtract 3π
   Find the sine of the result
   Find the arcsine of the result
   Add 3π

   The result should be 11 exactly.

1.	Start with 3	
	Divide by 2	1.5
	Tangent of the result	14.10
	Cosine of the result	0.03574

Arccosine of the result	1.535 is the result, but in a nested calculation 14.10 would be expected. But $14.1 = 1.535 + 4\pi$. The calculator returns the value that is in the range 0–π.

2. Start with 11

Subtract 3π	1.57522039
Find the sine of the result	0.999990206
Find the arcsine of the result	1.566370614

The value 1.57522039 above is very close to $\frac{\pi}{2}$. The rate of change of the sine function at $\frac{\pi}{2}$ is 0, therefore the arcsine function is highly sensitive to rounding errors in this region.

Summary of the main points

- Calculators are great but they have their limitations. What is the largest number your calculator can handle?
- There are two solutions for $\sin x$, $\cos x$ and $\tan x$ in the range $0 \leq x < 360°$; your calculator will give you only one of them.
- Trigonometric functions of very large angles may show small errors.
- Functions may show significant errors where they are insensitive, for example, $\cos^{-1} y$ or $\sin^{-1} y$ where $y \approx 1$, or $\tan x$ where $x \approx \frac{\pi}{2}$.
- The numbers e and π have their own buttons on the calculator, but because they are irrational, their values are approximate.

11 Bad notation

Sloppy writing leads to sloppy thinking, which leads to errors. Here are some examples of where things can go wrong, usually through carelessness and ambiguity.

Bad notation 1

Is it true that $-2^2 = 4$?

It depends on what is meant by the notation on the left-hand side of the equals sign. It could be ambiguous. Is the intention to find the square of –2 or is it to square 2 and then change the sign of the result?

There is an order to undertaking arithmetic and algebraic operations and the oft-used mnemonic for this is BODMAS which stands for Brackets, Of, Divide, Multiply, Add, Subtract. 'Of' is a rather redundant term since it is really the same as Multiply and is usually used with respect to fractions, for example, '$\frac{2}{3}$ of 3 is 2' is the same as $\frac{2}{3} \times 3 = 2$. However, 'Of' does at least put a vowel into something that would otherwise be difficult to pronounce. A better mnemonic is BEDMAS, where the E stands for exponent, that is, index or power.

Using BODMAS or BEDMAS we now know the order of operation should be to square the 2 first and then change the sign, so that $-2^2 = -4$ is the correct answer. Most calculators should give this answer. However, if the intention was really to find the square of –2 then the equation should have been written as $(-2)^2 = 4$. Notice that brackets have now been introduced; BEDMAS leads to the correct answer because the contents of the brackets must now be evaluated first.

Bad notation 2

How would you expand the brackets in this expression $4(\{x+2)^2+z\}$?

You can't, the brackets are incorrect because they don't match. The rule with BEDMAS (or BODMAS) is that when evaluating an expression you must deal with the part of the expression inside the brackets first. When there are nested brackets, that is, brackets inside other brackets, you deal with the inner brackets before the outer brackets; actually that is implied within the BEDMAS rule anyway.

When there are nested brackets, the brackets are often written in different styles to aid identification of the associated openers and closers. So if you start inside the deepest level of bracketing within an expression then, as you work outwards from the centre, the brackets will appear in their correct pairs. For example, if you evaluate the following, the steps would be:

$$\begin{aligned}2[13-3\times\{7-2\times(5-3)\}] &= 2[13-3\times\{7-2\times 2\}]\\ &= 2[13-3\times 3]\\ &= 2\times 4\\ &= 8\end{aligned}$$

However, $4(\{x+2)^2+z\}$ appears to have got its brackets crossed so that the brackets are not nested. Therefore, this expression as written is meaningless. If your calculator or software does not recognise the implied pairing of brackets and treats them as though they were all in the same style, then this expression would be evaluated as $4((x+2)^2+z)$, in which case the expansion would be:

$$\begin{aligned} 4\left((x+2)^2+z\right) &= 4\left(x^2+4x+4+z\right) \\ &= 4x^2+16x+16+4z \end{aligned}$$

Bad notation 3

Is this equation correctly evaluated?

$$4+2x-5x=4+-3x$$

Well, at the very least it is untidy. Actually, it is wrong to have more than one arithmetic operator operating on a variable, number or expression. What the writer has failed to do is to evaluate the result of the operations concerned. In this case, if you add a negative quantity the result is a negative quantity, so the correct evaluation should have been:

$$4+2x-5x=4-3x$$

Bad notation 4

Could this equation be better expressed?

$$y=x^2+x^1+x^0$$

While the expression may be perfectly valid there is a clear lack of understanding of what the terms mean. x^1 is the same as x; the exponent of 1 is redundant. $x^0=1$ whatever the value of x, so can replace it. Therefore, the equation could be more neatly written as:

$$y=x^2+x+1$$

Bad notation 5

What is poorly expressed in this equation?

$$(3x+2)(x-1)=3x^2-1x-2$$

The expansion of the brackets is correct but one of the terms is at the very least inelegant. In writing a term it is usual to lead with the numerical coefficient, so the $3x^2$ is correctly expressed. However, although the $1x$ starts with a numerical coefficient a numerical coefficient of 1 is redundant because multiplying anything by 1 does not change its value. The equation should have been written as:

$$(3x+2)(x-1)=3x^2-x-2$$

Bad notation 6

Is this equation correct?

$$2/3x=2x/3$$

It depends on what the writer meant. As written, the left-hand side of the equation is ambiguous. Does it mean $\frac{2}{3}$ of x or 2 divided by $3x$? If the former, the equation is correct. If the latter, it is not. How could the ambiguity be avoided? The problem lies in the fact that all the digits and symbols are written on the same line. Either fractions need to be written with explicit denominators and numerators or brackets are required. Therefore, the equation above could be written in either of these forms.

$$\frac{2}{3}x=\frac{2x}{3}$$

$$(2/3)x=2x/3$$

Bad notation 7

This expression

3^2^0.5

appears to show a power of a power, but what is its value? The 3^2 element of the expression is one way of writing 3 to the power of 2, that is, 3^2. This form of symbolism is used in some calculators and in Microsoft's spreadsheet, Excel. The ambiguity comes from deciding which of the following the above expression should be:

$$3\wedge(2\wedge 0.5) = 4.7288 \quad \text{or}$$

$$(3\wedge 2)\wedge 0.5 = 3$$

Unlike multiplication, where in $3 \times 2 \times 0.5$ it is immaterial in which order the multiplications are performed.

$$(3 \times 2) \times 0.5 = 3 \times (2 \times 0.5)$$

The order of operation with powers of powers is important. To avoid ambiguity brackets should be used as above. If brackets for 3^2^0.5 are not used in a Casio fx-85MS calculator or in Excel the result is 3. This implies that the order operation is $(3\wedge 2)\wedge 0.5$, that is, working from left to right across the expression. However, this is incompatible with the notation that if $3\wedge 2 = 3^2$ then $3\wedge 2\wedge 0.5 = 3^{2^{0.5}}$. So, if there is any chance of ambiguity, use brackets to make the meaning clear.

Bad notation 8

Multiplication signs are usually left out in algebraic expressions and their presence is implied by the absence of any other operator. So, for example, $2x$ means $2 \times x$. Likewise, if the multiplication signs are removed,

$$\ln(y^{-\beta}) \times 3 \times x \times \pi \times A \times \sin(p) \times h^4 = \ln(y^{-\beta})\, 3x\pi\, A \sin(p)\, h^4$$

However, the right-hand side of this equation, although not wrong, is messy. The usual convention in any product is that the numerical coefficient comes first, followed by constants, then single variables in alphabetical order, followed by expressions in increasing complexity. Therefore, the right-hand side of the above should be written like this:

$$= 3\,\pi\, A\, x\, h^4\, \sin(p)\, \ln(y^{-\beta})$$

Bad notation 9

Handwriting is often the enemy of neatness. What went wrong here?

$$\frac{x+y}{a+b} = {}^{x+y}\Big/_{a+b}$$
$$= x + y/a + b$$
$$= x + \frac{y}{a} + b$$

If you must use the / notation to indicate division, then it is necessary to make clear just what is being divided by what. On the left-hand side of the first line above it is $x+y$ that is being divided by $a+b$; that is clear enough. On the right-hand side of the first line it has become unclear whether it has the same meaning. On the second line it looks more like the only division taking place is that of is y by a. By the time we get to the last line that is indeed the case and therefore the opening and closing expressions are definitely not equal.

$$\frac{x+y}{a+b} \neq x + \frac{y}{a} + b$$

How could the ambiguity have been avoided? Answer: simply by using brackets:

$$\frac{(x+y)}{(a+b)} = {}^{(x+y)}\Big/_{(a+b)} = (x+y)/(a+b)$$

Bad notation 10

Sometimes when under pressure of time, such as during an examination, errors are made in intermediate calculations. What is wrong here?

$$\frac{d\left(x\left(x^2+1\right)\right)}{dx} = x^3 + x = 3x^2 + 1$$

The final answer is right, and the logical process is apparent; however, the middle part is clearly wrong. It should have been:

$$\frac{d\left(x^3+x\right)}{dx}$$

Bad notation 11

Is this equation correct?

$$\int x^2+x=\frac{x^3}{3}+\frac{x^2}{2}+C$$

It is not possible to say because some of the essential notation is missing. What is being integrated? x^2 or x^2+x? If it is just x^2 the answer is wrong; if x^2+x then it is right. So how should the problem have been expressed?

$$\int x^2+x\,dx$$

The missing part was the *dx*, which with the $\int$ symbol, indicates what is to be integrated. In fact, without the *dx* it is not even clear what the x^2+x is to be integrated with respect to.

Quiz

How can you improve upon the notation of the following?

1. $(3y-2)(2x-3)=6xy-4x-9y-6$
2. $\left(x^2+x^{-1}\right)\left(x^1+x^{-2}\right)=x^{(2+1)}+x^{(2-2)}+x^{(-1+1)}+x^{(-2-1)}$
   $=x^3+x^0+x^0+x^{-3}$
   $=x^3+2x^0+x^{-3}$
3. What are the errors in this example of indefinite integration?

$$\int e^x=e^x$$

4. Evaluate this fraction, giving your answer to 4 decimal places.

$$^{2/3}\Big/_{4/5}$$

1. $= 6xy - 4x - 9y + 6$
2. $= x^3 + 2 + x^{-3}$
3. There are two errors: one of notation and the other in the missing constant of integration. This should have been written as:

$$\int e^x dx = e^x + C$$

4. It depends on how you interpret this. The notation is highly ambiguous and it is unclear what is meant.

Is it $2 \div 3 \div 4 \div 5 = 0.0333$

or $2 \div 3/4 \div 5 = \dfrac{2 \times 4}{3 \times 5} = 0.5333$

or $2/3 \div 4/5 = \dfrac{2 \times 5}{3 \times 4} = 0.8333$?

Brackets should be used to indicate which sub-fractions should be computed first. They are essential if the expression is to be unambiguous.

Summary of the main points

- When negative numbers are to be raised to a power, the use of brackets may avoid ambiguity. What does -3^2 mean: $-(3^2) = -9$ or $(-3)^2 = 9$?

- Nested brackets must be inside each other; they may not overlap.
- There may be only one operator between any two expressions. For example, you cannot have $2+-5y$. Does that really mean $2-5y$?
- Spurious indices are often unhelpful. For example, $x^0+x^1+x^2$ would be better expressed as $1+x+x^2$.
- Multiplying by the numerical coefficient of 1 is, at best, unnecessary; $1x^2$ would be better as just x^2.
- Fractions are best shown in this format $\frac{1}{5x}$. The alternative format $1/5x$ is ambiguous because it could be interpreted as $\frac{1}{5x}$ or $\frac{1}{5}x$.
- The 'power of a power notation' in the form 3^16^0.5 is ambiguous. It could mean (3^16)^0.5 = 6561 or 3^(16^0.5) = 81. Use brackets to avoid the ambiguity.
- Products usually have the numerical coefficient first, followed by constants, single variables in alphabetical order, and then expressions in increasing order of complexity. For example, $= 5\,\pi\, abd^3\, \cos(z)\, \ln(\theta^3)$.
- Untidy handwriting may suggest muddled thinking; it often causes error.
- Make sure intermediate lines of calculation are fully stated. Shortcuts and omissions often lead to error.

12 Errors in calculus

Differentiation of powers

Calculus, like algebra, tends to be one of the more challenging subjects for science and engineering students of mathematics. As with other subjects, errors are sometimes made because of misunderstanding of the subject and sometimes because of carelessness.

Let's start by considering powers of a variable differentiated with respect to the variable.

Here is an easy mistake to spot.

$$\text{If} \quad y = 3x^4 \quad \text{then } \frac{dy}{dx} = 12x$$

The rule for differentiating the power of a variable runs like this:

> If $y = ax^n$, where a and n are constants and x and y are variables, then $\frac{dy}{dx} = anx^{n-1}$.

Therefore, if $y = 3x^4$ then $a = 3$ and $n = 4$, so $\frac{dy}{dx} = 12x^3$. What was missing in the differentiation at the top of this page was the index of 3.

Now consider this one:

$$\text{If } y = x^{\frac{1}{3}} \quad \text{then } \frac{dy}{dx} = \frac{1}{3}x^{-\frac{1}{3}}$$

At first glance this may appear correct, until you examine the index in the derivative. Here $n = \frac{1}{3}$ so $n - 1 = \frac{1}{3} - 1 = -\frac{2}{3}$. So the derivative should be:

$$\frac{dy}{dx} = \tfrac{1}{3}x^{-\frac{2}{3}}$$

Quiz

Which of the following are correct? In all cases, x and y are variables and n is a constant.

1. If $y = 2x^{2.3}$ then $\dfrac{dy}{dx} = 4.6x^{1.3}$
2. If $y = \pi x^{2\pi}$ then $\dfrac{dy}{dx} = \pi^2 x^{\pi}$
3. If $y = x^{n!}$ then $\dfrac{dy}{dx} = n!x^{(n-1)!}$
4. If $y = 2.5x^{\frac{2}{5}}$ then $\dfrac{dy}{dx} = x^{-\frac{3}{5}}$
5. If $y = -x^{-\frac{1}{3}}$ then $\dfrac{dy}{dx} = -\dfrac{1}{3}x^{-\frac{4}{3}}$
6. If $y = xn^{\frac{1}{3}}$ then $\dfrac{dy}{dx} = \dfrac{1}{3}xn^{-\frac{2}{3}}$

1. Correct.
2. No, $\dfrac{dy}{dx} = 2\pi^2 x^{(2\pi - 1)}$.
3. No, $\dfrac{dy}{dx} = n!x^{(n!-1)}$.
4. Correct.
5. No, $\dfrac{dy}{dx} = \dfrac{1}{3}x^{-\frac{4}{3}}$.
6. No, $\dfrac{dy}{dx} = n^{\frac{1}{3}}$ (x is the variable, n is a constant).

Differentiation of products

Here is a product of two expressions:

$$y = \left(2x^2 - x\right)\left(x^3 - x^2\right)$$

And therefore this is the derivative:

$$\frac{dy}{dx} = (4x - 1)\left(3x^2 - 2x\right)?$$

Here each expression has been differentiated and the product of those derivatives presented as the derivative of the original function. This is not correct.

There are two ways to find the derivative of y above. You could expand the brackets in the original function:

$$y = \left(2x^2 - x\right)\left(x^3 - x^2\right) = 2x^5 - 3x^4 + x^3$$

Differentiate term by term:

$$\frac{dy}{dx} = 10x^4 - 12x^3 + 3x^2$$

Alternatively use the product rule, which is that if $y = uv$, where u and v are themselves functions of x, then

$$\frac{dy}{dx} = u\frac{dv}{dx} + v\frac{du}{dx}.$$

In $y = \left(2x^2 - x\right)\left(x^3 - x^2\right), u = 2x^2 - x$

and $v = x^3 - x^2$, therefore

$$\frac{du}{dx} = 4x - 1 \quad \text{and} \quad \frac{dv}{dx} = 3x^2 - 2x$$

So $\dfrac{dy}{dx} = u\dfrac{dv}{dx} + v\dfrac{du}{dx}$

$$= \left(2x^2 - x\right)\left(3x^2 - 2x\right) + \left(x^3 - x^2\right)(4x - 1)$$

$$= 6x^4 - 7x^3 + 2x^2 + 4x^4 - 5x^3 + x^2$$

$$= 10x^4 - 12x^3 + 3x^2$$

Here is an example to show the rate at which the surface area of a cylinder changes with a change in the radius. If the surface area of a cylinder is given by:

$$A = 2\pi r\left(r + h\right)$$

Then:
Either expand the brackets to give:

$$A = 2\pi r^2 + 2\pi rh$$

Differentiate term by term:

$$\frac{dA}{dr} = 4\pi r + 2\pi h = 2\pi\left(2r + h\right)$$

Or use the product rule where $A = uv$ so $u = 2\pi r$ and $v = r + h$, therefore

$$\frac{du}{dr} = 2\pi \quad \text{and} \quad \frac{dv}{dr} = 1$$

So $\dfrac{dA}{dr} = u\dfrac{dv}{dr} + v\dfrac{du}{dr}$

$$= 2\pi r \times 1 + \left(r + h\right)2\pi$$

$$= 2\pi r + 2\pi r + 2\pi h$$

$$= 2\pi\left(2r + h\right)$$

Quiz

Which of the following are correct?

1. If $y = (1 + x^2)(1 - x^2)$ then $\dfrac{dy}{dx} = 4x^3$
2. If $y = x^e e^x$ then $\dfrac{dy}{dx} = x^e e^x (1 + ex)$
3. If $y = x^\pi \sin x$ then $\dfrac{dy}{dx} = x^\pi (x \cos x + \pi \sin x)$
4. If $y = \sin x \cos x$ then $\dfrac{dy}{dx} = \cos^2 x - \sin^2 x$
5. If $y = x^2 (\cos x + \sin x)$ then $\dfrac{dy}{dx} = (x^2 + 2x)(\sin x + \cos x)$

1. No, $\dfrac{dy}{dx} = -4x^3$.
2. No, $\dfrac{dy}{dx} = x^e e^x \left(1 + ex^{-1}\right)$.
3. No, $\dfrac{dy}{dx} = x^\pi \cos x + \pi x^{\pi-1} \sin x = x^{\pi-1}(x \cos x + \pi \sin x)$.
4. Correct.
5. No, $\dfrac{dy}{dx} = x^2 (\cos x - \sin x) + 2x(\sin x + \cos x) = \left(2x - x^2\right) \sin x + \left(2x + x^2\right) \cos x$.

Quotient rule

A product is where one expression is multiplied by another. To differentiate a product we use the product rule, so if the expressions are u and v, and both are functions of x when $y = uv$, then $\frac{dy}{dx} = u\frac{dv}{dx} + v\frac{du}{dx}$.

The u and v are interchangeable. For example, if $y = x^2 \sin x$ then u could be x^2 and v could be $\sin x$. In that case, $\frac{dy}{dx} = x^2 \cos x + \sin x \times 2x$.

On the other hand, if u and v were defined the other way around as $u = \sin x$ and $v = x^2$, then $\frac{dy}{dx} = \sin x \times 2x + x^2 \cos x$ which of course is exactly the same.

A quotient is where one expression is divided by another. When differentiating a quotient, the quotient rule is used. This rule says that if $y = \frac{u}{v}$ and u and v are both functions of x, then $\frac{dy}{dx} = \frac{v\frac{du}{dx} - u\frac{dv}{dx}}{v^2}$.

The numerator looks remarkably similar to the derivative in the product rule so we might expect the interchangeable property to apply.

$$\text{So if} \quad y = \frac{3x^3 - 7}{x^2 + 3}$$

then is it true that

$$\begin{aligned}\frac{dy}{dx} &= \frac{(3x^3 - 7)\,2x - (x^2 + 3)\,9x^2}{(x^2 + 3)^2} \\ &= \frac{6x^4 - 14x - 9x^4 - 27x^2}{(x^2 + 3)^2} \\ &= \frac{-3x^4 - 27x^2 - 14x}{(x^2 + 3)^2} \\ &= \frac{-x\,(3x^3 + 27x + 14)}{(x^2 + 3)^2}?\end{aligned}$$

Not so. Interchanging u with v in the numerator leads to a different answer for the derivative, $\frac{+x(3x^3+27x+14)}{(x^2+3)^2}$.

Why should the u and v be interchangeable in the product rule but not in the quotient rule? The answer is simply that in a product $uv = vu$ but in a quotient $\frac{u}{v} \neq \frac{v}{u}$.

Quiz

Find the derivatives with respect to x of the following quotients:

1. $y = \dfrac{x^2 - 2}{x^2 + 2}$
2. $y = \dfrac{\sin x}{\cos x}$
3. $y = \dfrac{e^x}{\ln x}$

1. $\dfrac{dy}{dx} = \dfrac{(x^2+2)2x-(x^2-2)2x}{(x^2+2)^2} = \dfrac{8x}{(x^2+2)^2}$

2. $\dfrac{dy}{dx} = \dfrac{\cos x \cos x - \sin x(-\sin x)}{\cos^2 x} = \dfrac{\cos^2 x + \sin^2 x}{\cos^2 x}$ $= \dfrac{1}{\cos^2 x} = \sec^2 x$. This is the answer you might have expected since $\dfrac{\sin x}{\cos x} = \tan x$.

3. $\dfrac{dy}{dx} = \dfrac{\ln x \times e^x - e^x \frac{1}{x}}{(\ln x)^2} = \dfrac{e^x(x \ln x - 1)}{x(\ln x)^2}$

Chain rule—missing links

When differentiating a function of a function we use the chain rule. The chain rule, at its simplest, says that if $y = f(u)$ and $u = g(x)$ then $\frac{dy}{dx} = \frac{dy}{du} \times \frac{du}{dx}$.

So, for example, if $y = \sin^2 x$ then $y = u^2$ where $u = \sin x$. In this case,

$$\frac{dy}{dx} = \frac{dy}{du} \times \frac{du}{dx} = 2u \times \cos x = 2 \sin x \cos x$$

If you are already familiar with the chain rule you might have got straight to the answer without the need for the intermediate working in the example above. If the function for y is more complicated, there may be a danger of missing part of the answer.

Check this out. If $y = \sqrt{x^2+1}$ then $\frac{dy}{dx} = \frac{1}{2\sqrt{x^2+1}}$?

Whenever square, or other, root signs appear in a calculus problem it is usually wise to re-express the function in exponent form. So the problem might be better written as $y = (x^2+1)^{\frac{1}{2}}$.

Now $y = u^{\frac{1}{2}}$ where $u = x^2+1$ so:

$$\frac{dy}{dx} = \frac{dy}{du} \times \frac{du}{dx} = \frac{1}{2}u^{-\frac{1}{2}} \times 2x = \frac{x}{\sqrt{u}} = \frac{x}{\sqrt{x^2+1}}$$

What was missing from the solution above was the $2x$; that is, the $\frac{du}{dx}$ term.

Where there is a function of a function of a function then the chain rule has three 'links' and might look like this:

$$\text{If} \quad y = f(u) \text{ and } u = g(v) \text{ and } v = k(x)$$

$$\text{then } \frac{dy}{dx} = \frac{dy}{du} \times \frac{du}{dv} \times \frac{dv}{dx}.$$

For example, to find the derivative of y where $y = e^{\sin^2 x}$, then $y = e^u$, where $u = v^2$, and $v = \sin x$, so:

$$\frac{dy}{dx} = \frac{dy}{du} \times \frac{du}{dv} \times \frac{dv}{dx} = e^u \times 2v \times \cos x$$

$$= e^{\sin^2 x} \times 2\sin x \times \cos x = 2e^{\sin^2 x} \sin x \cos x$$

Quiz

Find the derivatives with respect to x of the following functions:

1. $y = \cos(e^x)$
2. $y = \sin\left(e^{(x^2)}\right)$
3. $y = \tan(\ln(x^3))$

1. $\dfrac{dy}{dx} = -e^x \sin(e^x)$
2. $\dfrac{dy}{dx} = 2xe^{(x^2)} \cos\left(e^{(x^2)}\right)$
3. $\dfrac{dy}{dx} = 3x^2 \frac{1}{x^3} \sec^2(\ln(x^3)) = 3x^{-1} \sec^2(\ln(x^3))$

Chain rule–missing bits

It is easy to miss bits out when working with the chain rule. What has gone wrong here?

$$\text{If} \quad y = \sqrt[3]{1 - x^{-2}} \quad \text{then} \quad \frac{dy}{dx} = 2x\left(1 - x^{-2}\right)^{-\frac{2}{3}}$$

We can check this out by following the process described in the last section. First rewrite the problem as:

$$y = \left(1 - x^{-2}\right)^{\frac{1}{3}}$$

Now find the derivative of y where $y = (1 - x^{-2})^{\frac{1}{3}}$. So if $y = u^{\frac{1}{3}}$, where $u = 1 - x^{-2}$, then:

$$\frac{dy}{dx} = \frac{dy}{du} \times \frac{du}{dx} = \frac{1}{3}u^{-\frac{2}{3}}\left(-\left(-2x^{-3}\right)\right) = \frac{2}{3}x^{-3}\left(1 - x^{-2}\right)^{-\frac{2}{3}}$$

Quiz

Be careful not to miss any bits here when you find the derivatives with respect to x of the following functions:

1. $y = \left(e^{-e}\right)^{-x}$
2. $y = e^{\left(-e^{-x}\right)}$

   How quickly can you spot the errors in the next two questions?

3. If $y = 3\cos(x) - \sin(2x)$ then $\dfrac{dy}{dx} = -3\sin(x) - \cos(2x)$
4. If $y = x^{-2} + x^{-1} + x^0 + x^1 + x^2 + x^3$ then $\frac{dy}{dx} = -2x^{-3} - x^{-2} - x^{-1} + x^0 + 2x + 3x^2$

1. $y = \left(e^{-e}\right)^{-x} = e^{ex}$ so $\dfrac{dy}{dx} = ee^{ex} = e^{(ex+1)}$
2. $\dfrac{dy}{dx} = e^{\left(-e^{-x}\right)}\left(-\left(-e^{-x}\right)\right) = e^{\left(-e^{-x}\right)}e^{-x} = e^{-\left(x+e^{-x}\right)}$
3. $\dfrac{dy}{dx} = -3\sin(x) - 2\cos(2x)$
4. $\dfrac{dy}{dx} = -2x^{-3} - x^{-2} + 1 + 2x + 3x^2$

Constants and variables

If you want to differentiate a product then you must use the product rule.

So if $y = uv$, where u and v are functions of x, then $\frac{dy}{dx} = u\frac{dv}{dx} + v\frac{du}{dx}$.

So, for example, to differentiate:

$$y = Ax^3$$

$$\text{then} \quad \frac{dy}{dx} = 3Ax^2 + x^3?$$

The answer is no, however you look at it. That answer rather depends on what A is. If A is a constant then the answer is simply $\frac{dy}{dx} = 3Ax^2$ and it is not necessary to invoke the product rule. However, if A is a variable or another function, then it must be treated as such, and so the product rule applies.

$$\frac{dy}{dx} = A\frac{d\left(x^3\right)}{dx} + x^3\frac{dA}{dx}$$

$$= 3Ax^2 + x^3\frac{dA}{dx}$$

Check this out. If $y = e^x$ then $\frac{dy}{dx} = xe^{x-1}$.

At first glance it might look correct. However, it is not a variable raised to a constant power that is to be differentiated but a constant raised to a variable power. Had the problem been that of $y = x^e$ then $\frac{dy}{dx} = ex^{e-1}$ would be true. However, given the problem above, if $y = e^x$ then $\frac{dy}{dx} = e^x$.

In calculus it is essential to identify functions, variables and constants and treat them accordingly.

Quiz

In the following, find the derivative of y with respect to x, $\frac{dy}{dx}$. In each case, assume that:

a. k is a constant and then
b. k is a variable or another function of x.

1. $y = \sin kx$
2. $y = \dfrac{\ln kx}{x^3}$
3. $y = e^{kx}$

1.a. $\frac{dy}{dx} = k\cos kx$

1.b. This requires the use of the chain rule because $y = \sin kx$ is now a function of a product. If $y = \sin u$, where $u = kx$, then

$$\frac{dy}{dx} = \frac{dy}{du} \times \frac{du}{dx} = \cos u \times \left(k\frac{dx}{dx} + x\frac{dk}{dx}\right)$$

because u is a product and so the product rule applies

$$\frac{dy}{dx} = \cos kx \times \left(k + x\frac{dk}{dx}\right)$$

2.a. This is a quotient, so the quotient rule applies. If $y = \frac{u}{v}$ then $\frac{dy}{dx} = \frac{v\frac{du}{dx} - u\frac{dv}{dx}}{v^2}$.
Therefore in this problem, $u = \ln kx = \ln k + \ln x$ and $v = x^3$. k is a constant, so:

$$\begin{aligned}\frac{dy}{dx} &= \frac{x^3\frac{1}{x} - 3ux^2}{x^6}\\ &= \frac{x^2 - 3x^2\ln kx}{x^6}\\ &= \frac{1 - 3\ln kx}{x^4}\\ &= x^{-4}(1 - 3\ln kx)\end{aligned}$$

2.b. This is still a quotient, so again the quotient rule applies.

Therefore in this problem, $u = \ln kx = \ln k + \ln x$ and $v = x^3$. k is not a constant, so:

$$\frac{dy}{dx} = \frac{x^3\left(\frac{d}{dx}(\ln k) + \frac{1}{x}\right) - 3ux^2}{x^6}$$

$$= \frac{x^3\left(\frac{d}{dx}(\ln k) + \frac{1}{x}\right) - 3x^2 \ln kx}{x^6}$$

$$= x^{-3}\frac{d}{dx}(\ln k) + x^{-4} - 3x^{-4}\ln kx$$

3.a. $\dfrac{dy}{dx} = ke^{kx}$

3.b. Again, this requires the use of the chain rule because $y = e^{kx}$ is now a function of a product. If $y = e^u$, where $u = kx$, then

$$\frac{dy}{dx} = \frac{dy}{du} \times \frac{du}{dx} = e^u \times \left(k\frac{dx}{dx} + x\frac{dk}{dx}\right)$$

Note that u is a product and so the product rule applies

$$= e^{kx} \times \left(k + x\frac{dk}{dx}\right)$$

Derivatives of constants

How quickly can you spot the error here? If $y = Ax^3 + Bx^2 + C$, where A, B and C are constants then $\frac{dy}{dx} = 3Ax^2 + 2Bx + C$. The correct answer is of course $\frac{dy}{dx} = 3Ax^2 + 2Bx$.

The constant standing alone is often missed. Differentiate a constant and you get 0. Since by definition a constant does not change, its derivative must be zero. Follow the patterns in these derivatives.

$$\text{If} \quad y = x^5 + x^4 + x^3$$

$$\text{then} \quad \frac{dy}{dx} = 5x^4 + 4x^3 + 3x^2$$

$$\text{and} \quad \frac{d^2y}{dx^2} = 4 \times 5x^3 + 3 \times 4x^2 + 2 \times 3x^1$$

$$\text{And if} \quad y = x^{-3} + x^{-4} + x^{-5}$$

$$\text{then} \quad \frac{dy}{dx} = -3x^{-4} - 4x^{-5} - 5x^{-6}$$

$$\text{and} \quad \frac{d^2y}{dx^2} = 4 \times 3x^{-5} + 5 \times 4x^{-6} + 6 \times 5x^{-7}$$

In both cases a simple sequence of x to decreasing powers, when differentiated, gives another simple sequence. But what happens here to break the pattern in this sequence that connects the two above?

$$\text{If} \quad y = x^3 + x^2 + x^1 + x^0 + x^{-1} + x^{-2} + x^{-3}$$

$$\text{then} \quad \frac{dy}{dx} = 3x^2 + 2x^1 + 1 - x^{-2} - 2x^{-3} - 3x^{-4}$$

$$\text{and} \quad \frac{d^2y}{dx^2} = 6x + 2 + 2x^{-3} + 6x^{-4} + 12x^{-5}$$

In the expression for y above, which is a sequence of x to decreasing powers, there is a term x^0. However, since this is just the constant 1, it disappears when differentiated. When y is differentiated again there appears to be two missing terms in the sequence for $\frac{d^2y}{dx^2}$, that is terms in x^{-1} and x^{-2}; this is just because of successive differentiation of constants. So watch out for any constants in the quiz.

Quiz

Find the derivatives, $\frac{dy}{dx}$, of the following.

1. $y = x^7$
2. $y = x^e$
3. $y = x^\pi$
4. $y = e^x$
5. $y = \pi^x$
6. $y = e^\pi$

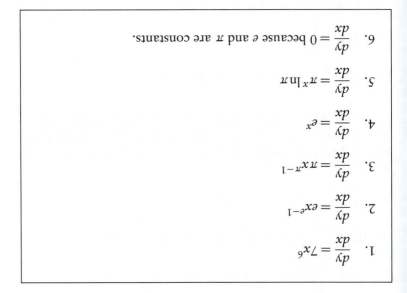

1. $\frac{dy}{dx} = 7x^6$
2. $\frac{dy}{dx} = ex^{e-1}$
3. $\frac{dy}{dx} = \pi x^{\pi-1}$
4. $\frac{dy}{dx} = e^x$
5. $\frac{dy}{dx} = \pi^x \ln \pi$
6. $\frac{dy}{dx} = 0$ because e and π are constants.

Patterns in derivatives of trigonometric functions

We know that if you differentiate $y = \sin(x)$ you get $\frac{dy}{dx} = \cos(x)$. So if you differentiate the related trigonometric function $y = \sec(x)$ then you must get $\frac{dy}{dx} = \csc(x)$?

Not so. Although successive differentiation of $y = \sin(x)$ leads to the pleasing cycle of derivatives

$$\frac{dy}{dx} = \cos(x), \quad \frac{d^2y}{dx^2} = -\sin(x), \quad \frac{d^3y}{dx^3} = -\cos(x), \quad \frac{d^4y}{dx^4} = \sin(x), \text{ etc.}$$

there is no such simple cycle for derivatives of $y = \sec(x)$.

$$\frac{dy}{dx} = \sec(x)\tan(x),$$

$$\frac{d^2y}{dx^2} = \sec^3(x) + \sec(x)\tan^2(x),$$

$$\frac{d^3y}{dx^3} = 3\sec^3(x)\tan(x) + 2\sec^2(x)\tan^2(x) + \sec(x)\tan^3(x)$$

and so on with increasing complexity.

Quiz

Which of the following, when differentiated four times, return to the original function? That is, for which of the following is it true that $\frac{d^4y}{dx^4} = y$?

1. $y = \cos(x)$
2. $y = \sin(x)$
3. $y = \tan(x)$
4. $y = e^x$
5. $y = e^{-x}$
6. $y = \ln x$

All are true except for numbers 3 and 6.

Inverse of a trigonometric function

If $y = \cos^{-1}(x)$ then, using the chain rule for differentiating a function of a function, does $\frac{dy}{dx} = \cos^{-2}(x)\sin(x)$?

This all depends on what is meant by $\cos^{-1}(x)$. Let's start by taking a step backwards to examine this notation and its meaning. $\cos(x)$ is the cosine of x, that is straightforward enough. $\cos(x)$ raised to a power, cubed for example, is written as $\cos^3(x)$ and means $(\cos(x))^3$. The reciprocal is written as $\cos^{-3}(x)$ and means $\frac{1}{(\cos(x))^3}$. So

$$\text{if} \quad \cos^{-3}(x) \text{ means } \frac{1}{(\cos(x))^3}$$

$$\text{then} \quad \cos^{-2}(x) \text{ means } \frac{1}{(\cos(x))^2}$$

and you might therefore expect that

$$\cos^{-1}(x) \quad \text{means} \quad \frac{1}{(\cos(x))}.$$

But that is not so. The –1 index has a very specific meaning and it does not indicate the reciprocal of the trigonometric function, but rather its inverse. So

$$\text{if} \quad x = \cos(y)$$

$$\text{then} \quad y = \cos^{-1}(x)$$

You will probably see the $\cos^{-1}(x)$ function above the $\cos(x)$ button on your calculator.

So how would you write $\frac{1}{(\cos(x))}$ as $\cos(x)$ raised to the power of –1? It would have to be in the rather more heavy-handed notation of $(\cos(x))^{-1}$ to avoid its being mistaken for the inverse function.

Back to the top line of this section, if $y = \cos^{-1}(x)$ then the chain rule has no place here for differentiating y because $\cos^{-1}(x)$

is not a function of a function. Rather, we can find the derivative of $\cos^{-1}(x)$ like this:

Start with the inverse function	$x = \cos(y)$
Differentiate with respect to y	$\dfrac{dx}{dy} = -\sin(y)$
Find the reciprocal of both sides	$\dfrac{dy}{dx} = -\dfrac{1}{\sin(y)}$
But $\sin(y) = \sqrt{1-\cos^2(y)}$, so	$\dfrac{dy}{dx} = -\dfrac{1}{\sqrt{1-\cos^2(y)}}$
And, since $\cos(y) = x$, substitute	$\dfrac{dy}{dx} = -\dfrac{1}{\sqrt{1-x^2}}$

Similar derivations can be used to show that $\frac{d}{dx}\left(\sin^{-1}(x)\right) = \frac{1}{\sqrt{1-x^2}}$ and $\frac{d}{dx}\left(\tan^{-1}(x)\right) = \frac{1}{1+x^2}$

Quiz

Can you find $\frac{dy}{dx}$ for each of the following without too much tortuous differentiation?

1. $y = \cos^{-1}(\cos(x))$
2. $y = \sin^{-1}(\cos(x))$
3. $y = \cos^{-1}(\tan(x))$

1. Here the inverse of a function is applied to the same function and so $y = x$. Therefore, $\frac{dy}{dx} = 1$.
2. $y = \sin^{-1}(\cos(x))$ is a function of a function So to differentiate it we need to use the chain rule. Rewrite $y = \sin^{-1}(\cos(x))$ as $y = \sin^{-1}(u)$, where $u = \cos(x)$.

The chain rule is: $\dfrac{dy}{dx} = \dfrac{dy}{du} \times \dfrac{du}{dx}$

$$\text{so} \quad \frac{dy}{dx} = \frac{1}{\sqrt{1-u^2}} \times (-\sin(x))$$

$$= \frac{1}{\sqrt{1-\cos^2(x)}} \times (-\sin(x))$$

$$= -\frac{1}{\sin(x)} \times \sin(x)$$

$$= -1$$

3. $y = \cos^{-1}(\tan(x))$ is another function of a function, so again we use the chain rule. Rewrite $y = \cos^{-1}(\tan(x))$ as $y = \cos^{-1}(u)$, where $u = \tan(x)$.

The chain rule is: $\dfrac{dy}{dx} = \dfrac{dy}{du} \times \dfrac{du}{dx}$

$$\text{so} \quad \frac{dy}{dx} = \frac{-1}{\sqrt{1-u^2}} \times \sec^2(x)$$

$$= \frac{-1}{\sqrt{1-\tan^2(x)}} \times \sec^2(x)$$

$$= \frac{-1}{\sqrt{\frac{\cos^2(x)}{\cos^2(x)} - \frac{\sin^2(x)}{\cos^2(x)}}} \times \sec^2(x)$$

$$= \frac{-\cos(x)}{\sqrt{\cos(2x)}} \times \sec^2(x)$$

$$= \frac{-1}{\cos(x)\sqrt{\cos(2x)}}$$

Hyperbolic and trigonometric functions

If $y = \sinh(x)$ then $\frac{dy}{dx} = \cosh(x)$, so if $y = \cosh(x)$ then does $\frac{dy}{dx} = -\sinh(x)$?

It is tempting to treat hyperbolic functions such as these in the same ways as trigonometric functions. After all, $y = \cosh(x)$ looks very much like $y = \cos(x)$. But these are very different functions. The x in $y = \cos(x)$ is usually taken to be an angle. The x in $y = \cosh(x)$ is far from an angle in the trigonometric sense.

$$\cosh(x) = \frac{e^x + e^{-x}}{2}$$

so, upon differentiation, $\dfrac{dy}{dx} = \dfrac{e^x - e^{-x}}{2} = \sinh(x)$.

So although $\dfrac{dy}{dx}(\sin(x)) = \cos(x)$ and $\dfrac{dy}{dx}(\cos(x)) = -\sin(x)$,

$$\frac{dy}{dx}(\sinh(x)) = \cosh(x) \text{ but } \frac{dy}{dx}(\cosh(x)) = \sinh(x)$$

When expressed as series, trigonometric and hyperbolic expansions do have some striking similarities.

$$\sin(x) = x - \frac{x^3}{3!} + \frac{x^5}{5!} - \frac{x^7}{7!} + \frac{x^9}{9!} - \ldots\text{and}$$

$$\sinh(x) = x + \frac{x^3}{3!} + \frac{x^5}{5!} + \frac{x^7}{7!} + \frac{x^9}{9!} + \ldots$$

$$\cos(x) = 1 - \frac{x^2}{2!} + \frac{x^4}{4!} - \frac{x^6}{6!} + \frac{x^8}{8!} - \ldots\text{and}$$

$$\cosh(x) = 1 + \frac{x^2}{2!} + \frac{x^4}{4!} + \frac{x^6}{6!} + \frac{x^8}{8!} + \ldots$$

$$\tan(x) = x + \frac{x^3}{3} + \frac{2x^5}{15} + \frac{17x^7}{315} + \frac{62x^9}{2835} + \ldots\text{and}$$

$$\tanh(x) = x - \frac{x^3}{3} + \frac{2x^5}{15} - \frac{17x^7}{315} + \frac{62x^9}{2835} - \ldots$$

Quiz

1. Can you use the series expansions for $\sinh(x)$ and $\cosh(x)$ above to derive a series expansion for e^x?
2. $\cosh(x) = \frac{e^x+e^{-x}}{2}$ and $\sinh(x) = \frac{e^x-e^{-x}}{2}$. Also, from trigonometry, we can get the equation $\cos^2(x) + \sin^2(x) = 1$. What similar looking equation relates $\sinh^2(x)$ and $\cosh^2(x)$?

1. $\cosh(x) + \sinh(x) = \frac{e^x+e^{-x}}{2} + \frac{e^x-e^{-x}}{2} = e^x$

Therefore,

$$e^x = \cosh(x) + \sinh(x) = 1 + \frac{x^2}{2!} + \frac{x^4}{4!} + \frac{x^6}{6!} + \frac{x^8}{8!} + \dots + x + \frac{x^3}{3!} + \frac{x^5}{5!} + \frac{x^7}{7!} + \frac{x^9}{9!} + \dots$$

$$= 1 + x + \frac{x^2}{2!} + \frac{x^3}{3!} + \frac{x^4}{4!} + \frac{x^5}{5!} + \frac{x^6}{6!} + \frac{x^7}{7!} + \frac{x^8}{8!} + \frac{x^9}{9!} + \dots$$

$$= \sum_{n=0}^{\infty} \frac{x^n}{n!}$$

2. $\sinh^2(x) = \left(\frac{e^x-e^{-x}}{2}\right)^2 = \frac{e^{2x}-2e^xe^{-x}+e^{-2x}}{4}$

$= \frac{e^{2x}-2+e^{-2x}}{4}$

and $\cosh^2(x) = \left(\frac{e^x+e^{-x}}{2}\right)^2 = \frac{e^{2x}+2e^xe^{-x}+e^{-2x}}{4}$

$= \frac{e^{2x}+2+e^{-2x}}{4}$

Therefore,

$\cosh^2(x) - \sinh^2(x) = \frac{e^{2x}+2+e^{-2x}-e^{2x}+2-e^{-2x}}{4} = 1$

Integration—missing bits

Integration is the inverse process to that of differentiation. Integration may be either indefinite or definite. In indefinite integration you are finding the anti-derivative of a function. That is, when the anti-derivative is differentiated you get your original function. For example, the anti-derivative of $3x^2$ is $x^3 + C$ (where C is an unknown constant) because the derivative of $x^3 + C$ is $3x^2$. So the indefinite integral is written as:

$$\int 3x^2 dx = x^3 + C$$

In definite integration you are doing the integration between chosen limits and so the constant of integration disappears because it is self-cancelling. For example:

$$\int_1^2 3x^2 dx = \left[x^3\right]_1^2 = 2^3 - 1^3 = 7$$

With indefinite integration the answer is normally an expression which includes a constant of integration, whereas with definite integration the result is normally a numerical value.

As with differentiation, it is easy to miss bits out if you are not careful.

Quiz

Can you see what is missing from or wrong with these integrals?

1. $\int \sin(x)\,dx = \cos(x) + C$
2. $\int_{-1}^{1} e^{-x} dx = [-e^{-x}]_{-1}^{1} = -e^{-1} - e^{-1} = 2e^{-1}$

3. $\displaystyle\int_{-1}^{1} x^2+x+1+x^{-1}+x^{-2}dx$

$$=\left[\frac{1}{3}x^3+\frac{1}{2}x^2+x^1-x^0-x^{-1}\right]_{-1}^{1}$$

$$=\frac{1}{3}+\frac{1}{2}+1-1-1-\left(-\frac{1}{3}+\frac{1}{2}-1+1+1\right)$$

$$=-\frac{4}{3}$$

1. $-\cos(x)+C$
2. $-e^{-1}-\left(-e^{-(-1)}\right)=e-e^{-1}$
3. $\left[\frac{1}{3}x^3+\frac{1}{2}x^2+x+\ln|x|-x^{-1}\right]_{-1}^{1}$

$$=\frac{1}{3}+\frac{1}{2}+1+0-1-\left(-\frac{1}{3}+\frac{1}{2}-1+0+1\right)$$

$$=\frac{2}{3}$$

Integration—missing constants

If you differentiate $\sin x$ with respect to x you get $\cos x$. Differentiate again with respect to x and you get $-\sin x$. Once more leads to $-\cos x$ and a fourth time leads back to $\sin x$. In conventional notation this could be expressed as:

$$\frac{d^4(\sin x)}{dx^4}=\sin x$$

So differentiate $\sin x$ with respect to x four times and you get $\sin x$ again. Since integration is the inverse process of differentiation then if you integrate $\sin x$ with respect to x four times you will get back to $\sin x$ again; right?

Not so. What is missing in this indefinite integration is the constant of integration. In fact, with four separate and successive integrations there will be four constants of integration which become the numerical coefficients of increasing powers of x. So

$$\int\left(\int\left(\int\left(\int(\sin x)dx\right)dx\right)dx\right)dx=\sin x+Ax^3+Bx^2+Cx+D$$

We can verify this by differentiating $\sin x+Ax^3+Bx^2+Cx+D$ four times:

$$\frac{d\left(\sin x+Ax^3+Bx^2+Cx+D\right)}{dx}=\cos x+3Ax^2+2Bx+C$$

$$\frac{d\left(\cos x+3Ax^2+2Bx+C\right)}{dx}=-\sin x+6Ax+2B$$

$$\frac{d(-\sin x+6Ax+2B)}{dx}=-\cos x+6A$$

$$\frac{d(-\cos x+6A)}{dx}=\sin x$$

So, don't forget the constants of integration in indefinite integration!

Quiz

Complete the following:

1. $\int\int x^n dx=\dfrac{x^{n+2}}{(n+1)(n+2)}+...$
2. $\int\left(\int\left(20x^3\right)dx+1\right)dx=...$
3. $\int\left(\int\left(x^2\right)dx\right)^2 dx=...$

1. $\frac{x^{n+2}}{(n+1)(n+2)} + Cx + D$

2. $\int\left(\int\left(20x^3\right)dx + 1\right)dx = \int\left(5x^4 + C + 1\right)dx = x^5 + (C+1)x + D$

   However, since C is an arbitrary constant this answer can be simplified to $x^5 + Cx + D$ by redefining C.

3. $\int\left(\int\left(x^2\right)dx\right)^2 dx = \int\left(\frac{x^3}{3} + C\right)^2 dx$

   $= \int\left(\frac{x^6}{9} + \frac{2C}{3}x^3 + C^2\right)dx = \frac{x^7}{63} + \frac{Cx^4}{6} + C^2x + D$

Making assumptions

In the following, an integration has been performed. x is a variable and n is a constant. There is an integral sign and a dx to indicate that we are integrating with respect to x and to show the function that is being integrated. On the right-hand side of the equals sign is the result which includes the constant of integration, C. It all looks right, so what could be wrong?

$$\int x^n dx = \frac{x^{n+1}}{n+1} + C$$

For the most part it is right, but it is not universally right. There is one exception that makes it not always true. Consider the case when $n = -1$. If the integration was correct, then the expression on the right-hand side of the integration sign would be

$$\frac{x^{-1+1}}{-1+1} + C = \frac{1}{0} + C, \quad \text{which is undefined.}$$

Evaluated between the limits of 1 and 2, say, this would give $\frac{1}{0} - \frac{1}{0}$, which is undefined. This is nonsense since the area under a

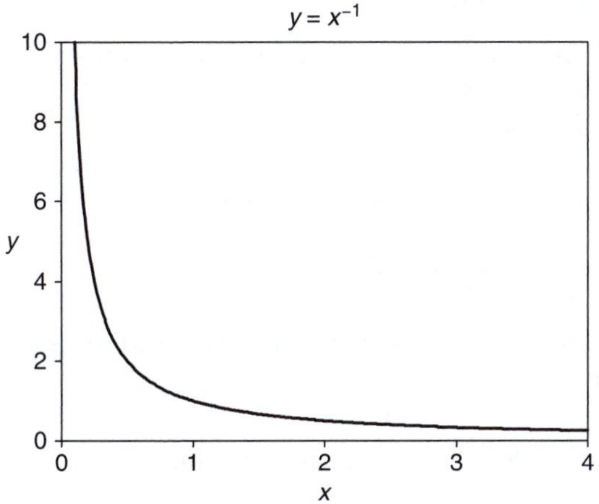

graph of x^{-1} could easily be estimated by the simple if inaccurate method of 'counting areas'.

The correct integral when $n = -1$ is:

$$\int x^{-1} dx = \ln|x| + C$$

This is the one exception to the generally correct rule for integrating x^n above.

Quiz

If $\int x^{-1} dx = \ln|x| + C$, what is $\int \ln|x|\, dx$?

$\int \ln|x|\, dx = x\ln|x| - x + C$, which may be confirmed by differentiating using the product rule:

$$\frac{d(x\ln|x| - x + C)}{dx} = x\frac{1}{x} + 1 \times \ln|x| - 1 = \ln|x|$$

Summary of the main points

- If $y = ax^n$, where a and n are constants and x and y are variables, then $\frac{dy}{dx} = anx^{n-1}$.
- The product rule is that:

  If $y = uv$, where u and v are functions of x,

  $$\text{then } \frac{dy}{dx} = u\frac{dv}{dx} + v\frac{du}{dx}.$$

- Terms need be identified as constants or variables. For example, if $y = zx^2$ then

  $$\frac{dy}{dx} = 2zx \text{ provided that } z \text{ is a constant, but}$$

  $$\frac{dy}{dx} = 2zx + x^2\frac{dz}{dx} \text{ if } z \text{ is a variable.}$$

- A constant differentiated with respect to a variable is 0. Constants are just that, constant, so they do not change.
- Successive differentiations of $\sin x$ or $\cos x$ just lead to similar terms. Differentiation of any other trigonometric function leads to successively more complicated expressions.
- Beware of differentiation of inverse trigonometric functions. They are as follows:

  If $y = \sin^{-1} x$ then $\frac{dy}{dx} = (1 - x^2)^{-\frac{1}{2}}$

  If $y = \cos^{-1} x$ then $\frac{dy}{dx} = -(1 - x^2)^{-\frac{1}{2}}$

  If $y = \tan^{-1} x$ then $\frac{dy}{dx} = (1 + x^2)^{-\frac{1}{2}}$

  If $y = \cot^{-1} x$ then $\frac{dy}{dx} = -(1 + x^2)^{-\frac{1}{2}}$

  If $y = \sec^{-1} x$ then $\frac{dy}{dx} = x^{-1}(x^2 - 1)^{-\frac{1}{2}}$

  If $y = \csc^{-1} x$ then $\frac{dy}{dx} = -x^{-1}(x^2 - 1)^{-\frac{1}{2}}$
- For these hyperbolic functions, $\frac{dy}{dx}(\sinh x) = \cosh x\,(x)$ and $\frac{dy}{dx}(\cosh x) = \sinh x$.

- The quotient rule is that:
  If $y = \frac{u}{v}$, where u and v are functions of x, then $\frac{dy}{dx} = \frac{v\frac{du}{dx} - u\frac{dv}{dx}}{v^2}$.
- The chain rule is used when differentiating a function of a function:

  If $y = f(u)$ and $u = g(x)$ then $\dfrac{dy}{dx} = \dfrac{dy}{du} \times \dfrac{du}{dx}$.

  If $y = f(u)$ and $u = g(v)$ and $v = k(x)$
  then $\dfrac{dy}{dx} = \dfrac{dy}{du} \times \dfrac{du}{dv} \times \dfrac{dv}{dx}$.

- With indefinite integration there is always a constant of integration, often written as C.
- With definite integration the constant of integration may be omitted because it becomes self-cancelling when you integrate the function between upper and lower limits.
- In successive integration you create a new constant of integration each time you perform the integration. For example, $\int\left(\int(x^5)\,dx\right)dx = \frac{x^7}{42} + Cx + D$.
- $\int x^n dx = \frac{x^{n+1}}{n+1} + C$ if $n \neq -1$. When $n = -1$ then $\int x^{-1} dx = \ln|x| + C$.

13 Test yourself

Here are 13 quizzes for you to test yourself with. Each one uses ideas developed in this book. Answers are given at the end of the chapter.

Q1

1. The potential energy of a mass is given by $PE = mgh$,where m is the mass, g is the acceleration due to gravity and h is the height of the mass above a reference surface. What are the units of potential energy in terms of kilograms, metres and seconds?
2. In a sale, successive discounts of 10%, 20%, 30% and 40% are applied. Does that mean you may claim the item for free?
3. Without using a calculator, divide $\frac{6}{35}$ by $\frac{9}{14}$.
4. Is this true that $a^{(b+c)}b^{(a+c)}c^{(a+b)} - (bc)^a (ac)^b (ab)^c = 1$?
5. Evaluate the expression $(-0.2)^6$ but without using a calculator.
6. Solve this quadratic equation: $\frac{x-4}{16-x^2} = 1$.

Need a hint?

For question 1 see chapter 3, question 2 – chapter 4, question 3 – chapter 7, questions 4, 5 and 6 – chapter 8.

Q2

1. The formula used to find the gravitational force of attraction, F, between two bodies is:

   $F = G\frac{m_1 m_2}{r^2}$, where G is the universal gravitational constant, m_1 and m_2 are the masses of the bodies and r is the distance between them. If F is in newtons, m_1 and m_2 are in kilograms and r is in metres, what are the units of G?

2. How many times does '20% off' have to be applied until the cost of an item is not more than 20% of its original cost?
3. Without using a calculator, divide $\frac{22}{39}$ by $\frac{121}{65}$.
4. Is this true that $\log(ab)^c = c\log(a)\log(b)$?
5. Find the real solution for the expression $\sqrt[15]{-1}$ but without using a calculator.
6. Evaluate this expression: $\log_x(y^2) \times \log_y(z^3) \times \log_z(x^4)$.

Need a hint?

For question 1 see chapter 3, question 2 – chapter 4, question 3 – chapter 7, questions 4, 5 and 6 – chapter 8.

Q3

1. An electron is believed to weigh 9.1×10^{-31} kilograms and have a radius of 10^{-22} metres. If spheres are packed together in three dimensions then the volume will consist of 26% of voids. What would a cubic metre of electrons weigh, assuming such a thing was possible?
2. Five coloured cars each have average fuel consumptions as follows:

   blue 51 mpg   blue 45 mpg   blue 57 mpg
   red 63 mpg   red 57 mpg

   If the third car is now painted red, how does that affect the average mpg of a red car and the average mpg of a blue car?

3. Without using a calculator, evaluate as a decimal $-(25)^{-\frac{3}{2}}$.
4. Is this true that $\log_e(10) \times \log_{10}(e) = 1$?
5. Is this correct? $\frac{3ab-5bc+7ba}{105abc} = \frac{1}{35c} - \frac{1}{21a} + \frac{1}{15b}$?
6. Which of the following are true: $(0 \times 1)^1 = 1^0$, $\infty^{-\infty} = -\infty$, $\frac{\infty}{-\infty} = -\infty$, $\infty^0 - 0^\infty = 1$?

Need a hint?

For question 1 see chapter 4, question 2 – chapter 4, question 3 – chapter 7, questions 4, 5 and 6 – chapter 8.

Q4

1. If all the eyeballs of all the people on earth today were laid side by side would the line stretch around the equator?
2. What connects these two expressions: $\sqrt[6]{\pi^5 + \pi^4}$ and $e^{\frac{87}{76}}$?
3. Factorise the expression $a^3b + 2a^2b^2 + ab^3$if possible.
4. If $\sqrt{x}$, $\sqrt{y}$ and $\sqrt{x+y}$ are all positive integers greater than zero, what are the smallest possible values for x and y?
5. Is this correct $\frac{2x+7xy+3xyz}{xz} = 2z + \frac{7y}{z} + 3y$?
6. Which mode is your calculator in if $\sin^{-1}(0.012) = 0.764$, $\sin^{-1}(0.012) = 0.012$ and $\sin^{-1}(0.012) = 0.688$?

Need a hint?

For question 1 see chapter 4, question 2 – chapter 5, questions 3, 4 and 5 – chapter 8, question 6 – chapter 9.

Q5

1. In the last general election a winning candidate gained 16 239 votes in a turnout of 48 196. What percentage of the poll was that, 33%, 33.69% or 33.6937%?
2. In one day I travel 243 miles, on the next 81 miles, on the next, 27 miles, on the next 9 miles, etc. How far could I possibly travel in total if I continue like this?
3. Simplify the expression $x\tan y + y\tan x$ if possible.

4. Simplify $y + (\sqrt{x} + \sqrt{y})(\sqrt{x} - \sqrt{y})$.
5. Is this correct $\frac{a^2}{a^3+a^4} = \frac{1}{a} + \frac{1}{a^2}$?
6. Without using a calculator, can you fill in the missing (?) trigonometric functions: $?(0°) = 1.0$, $?(45°) = 1.0$, $?(90°) = 1.0$?

Need a hint?
For question 1 see chapter 4, question 2 – chapter 5, questions 3, 4 and 5 – chapter 8, question 6 – chapter 9.

Q6

1. How long is a piece of string? You cannot be bothered to unroll the string, which is wound evenly around a cardboard tube. You estimate the diameter of the tube and the outside diameter of the string around the tube. You also estimate the number of turns from the length of the tube and the thickness of the string. How good will your estimate be?
2. Which of these statements are clearly untrue?

   a. Guns can kill you.
   b. Bullets can kill you.
   c. People can kill you.

3. Simplify the expression $(a+b)(b-a)(a^2+b^2)$ if possible.
4. What is the product of the square roots of –25 and –36?
5. How many real and distinct solutions are there for each of the following equations?

   a. $x^2 + 6x + 9 = 0$
   b. $2x^2 + 6x + 3 = 0$
   c. $3x^2 + 2x + 6 = 0$

6. How many real solutions for x are there to the following expression, where a is a constant and $-1 < a < 1$?

$$a = \cot^2 x, \quad \text{where } 0° \leqslant x < 360°$$

Need a hint?

For question 1 see chapter 4, question 2 – chapter 6, questions 3, 4 and 5 – chapter 8, question 6 – chapter 9.

Q7

1. You wish to buy a second-hand car for £5000 but cannot afford to do so without a loan. You are offered three possible deals:

   a. Pay £2000 deposit and borrow £3000 at 5% per month.
   b. Pay £1000 deposit and borrow £4000 at 4% per month.
   c. Pay no deposit and borrow £5000 at 3% per month.

   All the deals require you to pay back at the rate of £200 per month. Which deal should you take?
2. A bridge is shown on three different maps at scales of 1:5000, 1:1000 and 1:2500. On which map does the bridge appear longest at the mapping scale?
3. Can you solve this equation $\frac{p+q-r+s-t}{t-s+r-q+2p} = -1$?
4. Express in its simplest form $(1+i)^4$.
5. Is this correct?

$$a^3 - 2a^2 - 5a + 6 = (a-3)(a+2)(a-1) = 0$$

$$\text{so } a = -3 \text{ or } a = 2 \text{ or } a = -1$$

6. Why cannot triangle ABC exist if $A = 137.0°$, $a = 3.72$ m and $C = 44.0°$?

Need a hint?

For question 1 see chapter 4, question 2 – chapter 7, questions 3, 4 and 5 – chapter 8, question 6 – chapter 9.

Q8

1. Of the following data set, which is the greatest, the mean, the mode or the median?

   6, 6, 6, 4, 3, 8, 9, 1, 11

2. The area of a field appears as 1000 square centimetres on a map of scale 1:1000. What will be the area of the same field on a map of scale 1:2500?
3. Can you solve this equation: $\frac{2}{3}(9a+27b+5c) = \frac{3}{2}(4a+12b-3)+5c$?
4. Without a calculator, evaluate $\left(\sqrt[4]{81}\times\sqrt[7]{128}\right)^3$.
5. Is this correct $\sqrt{\left(\sqrt[3]{a^2}\right)^3} = a$?
6. Is this true that $\int e^x = e^x$.

Need a hint?

For question 1 see chapter 6, question 2 – chapter 7, questions 3, 4 and 5 – chapter 8, question 6 – chapter 11.

Q9

1. What are the mean, mode and median of the following set of wine grape types?

Barbarossa	Black Muscat	Cabernet Sauvignon
Chardonnay	Chenin Blanc	Gamay
Lambrusco	Lazki Rizling	Merlot
Müller-Thurgau	Muscadet	Pignerol
Pinot Grigio	Pinot Noir	Riesling
Sauvignon Blanc	Sémillon	Shiraz
Tokay	Verdelho	Zinfandel

2. A nautical chart has a scale of 1:50 000. The width of a river estuary measured on the chart is 143 mm. How wide is the river in nautical miles?
3. Evaluate $(2^2)^2 - 2^{(2^2)}$.
4. Which is the correct answer for $\sqrt[4]{a^8b^{12}}$; a^4b^8, a^2b^4 or ab^2 ?
5. Is the following correct: If $f(a)=a^2-2$ then $f(2a+3)=4a^2+12a+7$?
6. Is it true that if $y=xn^{\frac{2}{5}}$ then $\frac{dy}{dx}=\frac{2}{5}xn^{-\frac{3}{5}}$?

Need a hint?

For question 1 see chapter 4, question 2 – chapter 7, questions 3, 4 and 5 – chapter 8, question 6 – chapter 12.

Q10

1. What are the RMS, arithmetic mean, geometric mean and harmonic mean of this data set: 2, 3, 4 and 5?
2. Evaluate $(-3)^2 - (-2)^3$.
3. Evaluate $\left(2^1\right)^0 - 2^{\left(1^0\right)}$.
4. Evaluate $(20 + 0.02) \times (20 + 0.02)$ without using a calculator.
5. If $f(a) = -a^2$ and $g(b) = b - b^2$, does $f(g(c)) = c^2\,(c-1)^2$?
6. Is it true that if $y = e^x x^e$ then $\frac{dy}{dx} = x^e e^x \left(1 + ex^{-1}\right)$?

Need a hint?

For question 1 see chapter 4, question 2 – chapter 7, questions 3, 4 and 5 – chapter 8, question 6 – chapter 12.

Q11

1. Which of the following 'averages': RMS, arithmetic mean, geometric mean and harmonic mean, would be most useful to find if you sought the average deviation in the speed of a driver who was attempting to maintain a constant speed?
2. Evaluate $\frac{-7^3}{(-7)^3}$.
3. Evaluate $\left(5^3\right)^1 - 1^{\left(3^5\right)}$.
4. Evaluate $(3\frac{1}{6})^2$ as a mixed number but without using a calculator.
5. Where is the error in this: $\frac{m^n}{n^m} + \frac{n^n}{m^m} = 1$ so $m^{mn} + n^{nm} = m^m n^m$?
6. Is it true that $\frac{d^{32}}{dx^{32}}(\cos y) = \cos y$?

Need a hint?

For question 1 see chapter 4, question 2 – chapter 7, questions 3, 4 and 5 – chapter 8, question 6 – chapter 12.

Q12

1. Which of the following 'averages': RMS, arithmetic mean, geometric mean and harmonic mean would be most useful to find if you wanted the average length of similar-sized pieces of wood?
2. Evaluate $(-1)^{321}$.
3. Is this true that $a^{(b-c)}a^{(c-b)} = 1$?
4. Find the real solution for this expression: $(-1)^{521}$.
5. Is the following correct: If $u^{-1} > v$ then $v^{-1} > u$?
6. If $y = 5\cos(x) - \sin(3x) + 4\tan(x)$ does $\frac{dy}{dx} = -5\sin(x) - \cos(3x) + \frac{4}{\cos^2 x}$?

Need a hint?

For question 1 see chapter 4, question 2 – chapter 7, questions 3, 4 and 5 – chapter 8, question 6 – chapter 12.

Q13

1. In the last election one candidate gained 10 698 votes. This represented an increase of 4.7%. How many votes did he get in the previous election?
2. Without using a calculator, evaluate as a fraction $(-243)^{-\frac{3}{5}}$.
3. Simplify the expression $\frac{\sin y}{x} + \frac{\sin x}{y}$ if possible.
4. Express in its simplest form $\frac{(a+ib)}{(a-ib)} - \left(\frac{a^2-b^2}{a^2+b^2}\right) - i\left(\frac{2ab}{a^2+b^2}\right)$.
5. Can you see what is missing from or wrong with this integral?

$$\int_{-1}^{1} -e^{-x}dx = [-e^{-x}]_{-1}^{1} = -e^{-1} + e^{1} = e - e^{-1}$$

6. $\int \left(\int (x^3)dx\right)^2 dx = \int \left(\frac{x^4}{4} + C\right)^2 dx = \int \left(\frac{x^8}{16} + C\frac{x^4}{2} + C^2\right)dx = \frac{x^9}{144} + C\frac{x^5}{10} + C^2x + D$, true or false?

Need a hint?

For question 1 see chapter 4, question 2 – chapter 7, questions 3 and 4 – chapter 8, questions 5 and 6 – chapter 12.

Answers

Answers to Q1

1. kg m^2 s^{-2}
2. No, the cost will be $0.9 \times 0.8 \times 0.7 \times 0.6 \times 100\% = 30.24\%$ of the original price.
3. $\frac{4}{15}$
4. No, it equals 0.
5. 0.000064
6. $x = -5$ only; $x = 4$ is not a solution because that leads to $\frac{0}{0} = 1$.

Answers to Q2

1. N m^2 kg^{-2}, which is m^3 kg^{-1} s^{-2}.
2. Eight times
3. $\frac{10}{33}$
4. No, $\log(ab)^c = c\log(ab)$.
5. –1
6. 24

Answers to Q3

1. The volume, v_e, of one electron is $\frac{4}{3}\pi\left(10^{-22}\right)^3 = 4.2 \times 10^{-66}\text{m}^3$.
   The number of electrons, n_e, you could get into 1 m^3 would be $n_e = (1 - 0.26)v_e^{-1} = 0.74 \times \left(4.2 \times 10^{-66}\right)^{-1} = 1.8 \times 10^{65}$.
   The mass of such a number of electrons would be $1.8 \times 10^{65} \times 9.1 \times 10^{-31} = 1.6 \times 10^{35}$ kg. This is about 800,000 times the mass of the sun.
   Such a situation is not possible (on earth) because electrons cannot be packed together as described. The spaces between

them and other sub-atomic particles are, relatively speaking, enormous.

2. Both averages fall; blue from 51 to 48 mpg and red from 60 mpg to 59 mpg.
3. –0.008
4. Yes
5. No, $\frac{3ab-5bc+7ba}{105abc} = \frac{10ab-5bc}{105abc} = \frac{2}{21c} - \frac{1}{21a}$.
6. None of them are true; $(0 \times 1)^1 = 0$ and $1^0 = 1$; $\infty^{-\infty}$, $\frac{\infty}{-\infty}$, ∞^0 and 0^∞ are all undefined and therefore the expressions using them cannot be evaluated.

Answers to Q4

1. The human eyeball is approximately 24 mm in diameter. There are about 6.8 billion people alive today and almost all have two eyes. Eyeball to eyeball that would extend to 326 400 kilometres. The circumference of the earth is just over 40 000 kilometres so all the world's human eyeballs would go about eight times around the earth.
2. The first is an approximation for e in terms of π and the second is an approximation for π in terms of e.
3. $ab(a+b)^2$
4. 9 and 16 (either way round)
5. No, $\frac{2x+7xy+3xyz}{xz} = \frac{2+7y}{z} + 3y$.
6. Grad, radian and degree modes, respectively.

Answers to Q5

1. Both 33% and 33.69% are meaningful statements; 33.6937% is not meaningful because at six significant figures it suggests that there is precision to 0.05 of a voter.
2. 364.5 miles
3. No simplification possible.
4. x
5. No, $\frac{a^2}{a^3+a^4} = \frac{1}{a+a^2}$.

6. Cosine or secant, tangent or cotangent, and sine or cosecant, respectively.

Answers to Q6

1. It all depends upon how well you can estimate the average diameter of a turn, the number of turns and the length of the tube. The estimate of the thickness of the string will affect the number of turns per centimetre run along the tube and also the number of layers of string. In practice this is probably the most inaccurate measurement, perhaps only good to about $\frac{3}{4}$ of its true value. As the error in the estimate of the thickness of the string will have the same effect on the estimate of the number of layers and the estimate of the number of turns per centimetre run, then along with the other sources of error this will lead to a solution for the length of the piece of string that is likely to be in the region of 50% in error.
2. a. is unlikely to be true unless you are hit with one; b. and c. could be true.
3. $b^4 - a^4$
4. –30. You cannot find this as the square root of the product of the square roots. $\sqrt{x}\sqrt{y} = \sqrt{xy}$ only if x and y are not both negative.
5. a. 1, b. 2, c. 0
6. 4 if $a > 0$, 2 if $a = 0$ and none if $a < 0$.

Answers to Q7

1. Deal a. is the best; you will pay off the loan in 29 months with a total payment of £7684 including deposit. Deal b. takes 42 months and costs £9207, while deal c. costs £9380 and takes 47 months.
2. 1:1000
3. $p = 0$ but q, r, s and t are indeterminate.
4. –4
5. No. $a = 3$ or $a = -2$ or $a = 1$.
6. The two given angles already add up to more than 180°.

Answers to Q8

1. They are all the same: 6.
2. 160 square centimetres
3. $c = 2.7$ but a and b are indeterminate.
4. 216
5. Yes
6. No, but $\int e^x dx = e^x + C$ is true.

Answers to Q9

1. Who could possibly say? This data set cannot be quantified or ranked as it stands. If some characteristic of the grapes, with a numerical value, were assigned to each name then it might be possible to find a mean, mode or median of the set of numbers; but not as the list stands.
2. 3.86 nautical miles
3. 0
4. None of these; it is a^2b^3.
5. Yes
6. No, x is the variable, not n. If $y = xn^{\frac{2}{5}}$ then $\frac{dy}{dx} = n^{\frac{2}{5}}$.

Answers to Q10

1. RMS 3.67, arithmetic mean 3.50, geometric mean 3.31, harmonic mean 3.12.
2. 17
3. –1
4. 400.8004
5. No. It is $-c^2(1-c)^2$.
6. Yes

Answers to Q11

1. Root mean squared (RMS)
2. 1
3. 124

4. $10\frac{1}{36}$
5. $\frac{m^n}{n^m} + \frac{n^n}{m^m} = 1$ becomes $m^{m+n} + n^{m+n} = m^m n^m$.
6. Yes

Answers to Q12

1. Arithmetic mean
2. –1
3. Yes
4. –1
5. Yes
6. No. $\frac{dy}{dx} = -5\sin(x) - 3\cos(3x) + \frac{4}{\cos^2 x}$.

Answers to Q13

1. The number of votes can be estimated as $\frac{10\,698}{1.047} = 10\,218$ to the nearest vote. Actually, the range of possible votes lies between 10 213 and 10 222 because the increase of 4.7% is only quoted to two significant figures.
2. $-\frac{1}{27}$
3. No simplification possible.
4. 0
5. $\int_{-1}^{1} -e^{-x} dx = \left[e^{-x}\right]_{-1}^{1} = e^{-1} - e^{1} = e^{-1} - e$
6. True

Glossary

Analogue Any value which is continuous is said to be analogue. A clock on which the hands move at a constant speed is analogue. If the hands jump every second or minute this is not analogue but discrete. Devices with digital displays are not analogue. In theory the indicator, for example, the hand, of an analogue instrument can show an infinite number of readings but in practice that number is restricted by physical limitations.

Axis (axes) In the Cartesian coordinate system the position of a point may be expressed in terms of its coordinates in two-dimensional space. The x- and y-coordinates of a point are, respectively, the distances of the point from the mutually perpendicular y- and x-axes, which meet at the origin of the system. In three dimensions the x-, y- and z-coordinates are respectively, the distances of the point from the mutually perpendicular yz-, xz- and xy-planes which intersect at the origin of the system.

Calculus Calculus is concerned with how functions change. This branch of mathematics consists of the two branches of differentiation and integration, otherwise known, respectively, as differential calculus and integral calculus. See **differentiation** and **integral.**

Coefficient of friction The coefficient of friction is the ratio of the magnitude of the force opposing sliding motion of a body, to the force between the body and the surface over which it slides. It is usually given the symbol μ and is a dimensionless quantity. The value of μ depends on the roughness of the body and of the surface over which the body slides.

Compound angle A compound angle is an angle which is the sum of more than one part.

Correlation Correlation is the dependence of one variable upon another. If there is complete positive dependence between the variables then there is a correlation of +1. If the dependence is completely negative then it is –1. A correlation of 0 indicates that the variables are completely independent. If there are two variables, x and y, with variances of σ_x^2 and σ_y^2 and the covariance between them is σ_{xy}, then the correlation is given by $\frac{\sigma_{xy}}{\sigma_x \sigma_y}$. The correlation statistic is unitless.

Decimal places When a number is quoted to so many decimal places that means it is to the prescribed number of digits after the decimal point. The number 1.234567 quoted to 2 decimal places would be 1.23 because there are 2 digits after the decimal point. The number 1.234567 to 3 decimal places would be 1.235, since 1.235 is closer to 1.234567 than 1.234 is. This is called rounding.

Differentiation Differentiation is concerned with finding the derivative of a function with respect to a given variable. That is, finding how one variable changes with respect to another variable. If the function may be represented graphically then differentiation can be used to find the slope of the graph. In mathematical notation, if $y = f(x)$, that is, y is a function of x, then the rate of change of y with respect to x may be written as y' or as $\frac{dy}{dx}$.

Discontinuity A discontinuity is a break in an otherwise continuous function.

For example, $y = \frac{1}{x}$ has a discontinuity when $x = 0$. As x tends to 0 from the negative direction y tends to $-\infty$ but as x tends to 0 from the positive direction y tends to $+\infty$.

Likewise, $y = \tan x$ has a discontinuity when $x = \frac{\pi}{2}$. As x tends to $\frac{\pi}{2}$ from less than $\frac{\pi}{2}$, $\tan x$ tends to $+\infty$ but as x tends to $\frac{\pi}{2}$ from greater than $\frac{\pi}{2}$, $\tan x$ tends to $-\infty$.

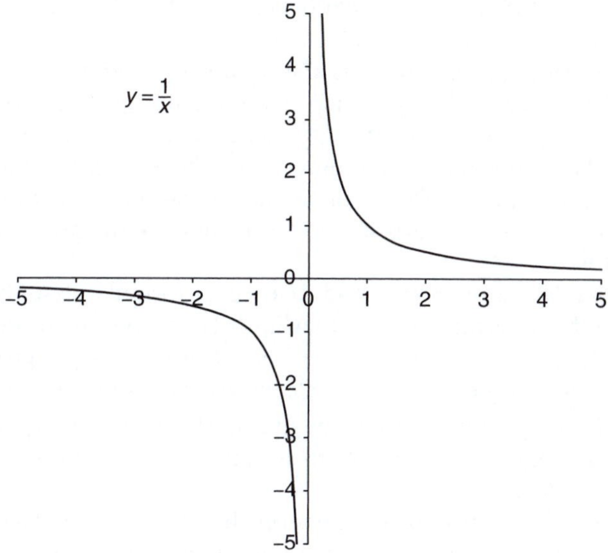

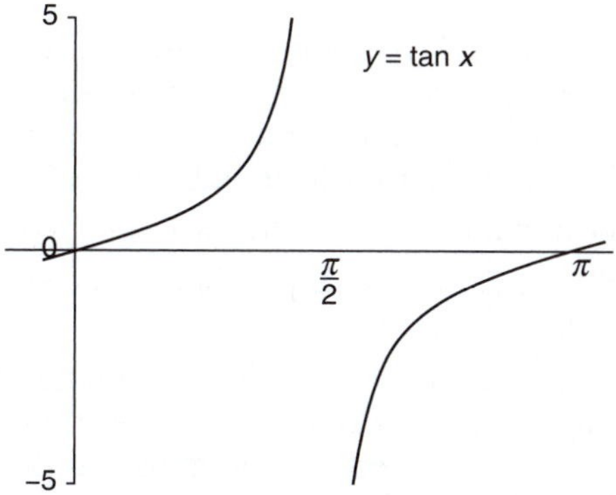

Factorial The factorial of a number is the product of all the positive integers from 1 to the number concerned. In notation it is written with an exclamation mark, !, after the number. So $1! = 1$, $2! = 1 \times 2 = 2$, $3! = 1 \times 2 \times 3 = 6$, $4! = 1 \times 2 \times 3 \times 4 = 24$ and so on; thus $5! = 120$, $6! = 720$, $7! = 5040$, $8! = 40320$,

$9! = 362880$, $10! = 3628800$, etc. Therefore, $n \times (n-1)! = n!$ and $n! \times (n+1) = (n+1)!$.

Factorisation Factorisation is the process of finding the factors of a number. The product of the factors gives the original number. The factors of 12 could be 2 and 6 or could be 3 and 4. The prime factors of a number are the set of prime numbers whose product is the original number. The prime factors of 12 are 2, 2 and 3 (not just 2 and 3 as they are the prime factors of 6).

Formula In mathematics and science a formula describes, in symbolic terms, the relationship between two or more variables. For example, $c = 2\pi r$ is the formula that gives the circumference, c, of a circle given its radius, r, and $a = \sqrt{b^2 + c^2}$ is the formula that gives the length of the hypotenuse of a right-angled triangle, a, given the lengths of the other two sides, b and c.

Function A function is an equation that relates an independent variable to a dependant variable. It is usually expressed in terms of inputs and outputs. See the **Functions** section in Chapter 8.

Hyperbolic functions Hyperbolic functions are analogous to trigonometric functions in that just as sin (sine) and cos (cosine) relate to a circle, $x^2 + y^2 = 1$, the counterparts of sinh and cosh relate to a hyperbola $x^2 - y^2 = 1$. The basic hyperbolic functions are: $\sinh x = \frac{e^x - e^{-x}}{2}$, $\cosh x = \frac{e^x + e^{-x}}{2}$, $\tanh x = \frac{e^{2x} - 1}{e^{2x} + 1}$. The terms are pronounced shine, cosh and than (th as in thin).

Hypothesis A hypothesis is a possible explanation of something that has not been or cannot be demonstrated to be true. Research often starts with a hypothesis and then investigates its likelihood. Research can never prove a hypothesis, only produce (statistical) evidence to support or reject it by evaluating its likelihood or otherwise. The phrase 'statistics prove that' is therefore a contradiction in terms. A conjecture on the other hand may be capable of proof but as yet that has not happened.

Imperial system The imperial system of units is the system of weights and measures established in the days of the former British Empire. Length includes units of inch, foot, yard and mile. Mass includes units of ounce, pound, stone, hundredweight and ton. Volume includes units of fluid ounce, pint, quart and gallon.

Inequality An inequality is a mathematical statement about the comparative sizes of two unequal quantities. The main inequality signs are: $\neq$ does not equal, $>$ greater than, $<$ less than, $\geqslant$ greater than or equal to and $\leqslant$ less than or equal to. See the **Inequalities** section in Chapter 8.

Infinity Infinity is an unbounded quantity that is often mistaken as a number. If infinity was a number then you could add 1 to it and so get a number beyond infinity, which is a contradiction in terms. Infinity is sometimes thought of as being the direction in which increasing positive numbers are heading, and minus infinity the direction of decreasing negative numbers. The symbol for infinity is ∞. Infinity is a tricky beast to deal with in mathematics and, like a caged hungry lion, needs to be handled with caution and respect. See also the **Zero and infinity** section in Chapter 8.

Integer Integers are positive or negative whole numbers. For example, $-7, 0, 2$ and 76723 are all integers; 1.2 and $\frac{3}{4}$ are not.

Integral An integral is the anti-derivative of a function. Given a derivative, an integral is the original function from which it is derived. So, for example, if you differentiate $x^3 + C$, where C is a constant, with respect to x, you get $3x^2$. Therefore, if you integrate $3x^2$ with respect to x you get $x^3 + C$.

Logarithm If a number can be expressed as a base raised to a power, then the power is the logarithm to the given base of the original number. Since $100 = 10^2$ then the logarithm of 100 to the base of 10 is 2. This may be written as $\log_{10} 100 = 2$. See **The logarithm function** section in Chapter 8.

Log table A log table is a tabulation of logarithms. In the pre-computer or pre-calculator era, when a computation requiring greater precision than could be achieved with a slide rule was to be done, then log tables would have been used. If two numbers were to be multiplied then the logarithms of the two numbers would be looked up and their values added together by hand. Then a table of anti-logs would be used to convert the sum just undertaken to find the product of the original two numbers. See also **slide rule** and **logarithm**.

Longitude The latitude and longitude of a point are often taken to be its geographical coordinates. Latitude is measured in angular terms from the equator in the direction of the North Pole and longitude is measured around the earth from the Greenwich meridian, positively eastwards. A meridian is a line from pole to pole passing through a point. More strictly, the latitude of a point is the angle that the vertical line through the point makes with the equatorial plane and longitude is the angle measured between two meridian planes, one of which is parallel to the vertical at the point and the other which is parallel to the vertical at Greenwich.

Mechanics Strictly this is a part of physics rather than mathematics, and is concerned with the application of forces on bodies and their displacement.

Metric system The metric system of units is the system of weights and measures established in 1791 in France. It has become the standard system in most countries, including the UK. Length is based on the metre with subdivisions and multiples of 10^3 as in micrometre, millimetre, metre and kilometre. Mass is based on the kilogram with subdivisions and multiples of 10^3 as in microgram, milligram, gram, kilogram and tonne. Volume is based on the litre with subdivisions and multiples of 10^3 as in millilitre, litre and kilolitre, although this last one is rarely used.

Parameter A parameter is a term that, while not being a variable, influences the outcome of an equation. For example, $y = mx + c$ is the equation of a straight line. Changing the value of the independent variable, x, will change the value of the dependent variable, y, for a given line defined by the slope, m, and the intercept on the y-axis, c. Since $y = mx + c$ represents all possible straight lines, then m and c are parameters of the equation of any given straight line.

Partial differential A partial differential, or partial derivative, is found by differentiating a function of more than one variable but with all but one of the variables being treated as constants. The notation is usually of the form of $f'x$ or as $\frac{\partial f}{\partial x}$.

Per cent Per cent, from Latin, means each hundred. A percentage is the equivalent decimal amount multiplied by 100 and is shown with a % sign. For example, 0.35 of a pile of 200 bricks is $0.35 \times 200 = 70$ bricks so 35% of 200 bricks is 35% of $200 = 70$.

Polynomial A polynomial is an expression which contains constants and variables with positive integer exponents all linked with the operations of add, subtract and multiply. For example, $2 - 3x + 4x^2 - 5x^3$ is a polynomial but $2 - 3x^{-1} + 4x^{\pi} - 5x^{\frac{3}{2}}$ is not a polynomial because it contains negative, non-integer and fractional exponents.

Power A power, also known as an index or an exponent, indicates the number of times a base must be multiplied by itself. For example, in the expression 2^3, 3 is the power and it means that the base, 2, must be multiplied by itself three times. So $2^3 = 2 \times 2 \times 2 = 8$. Likewise in algebraic terms, a^b means that the base a is multiplied by itself b times.

Product A product is the result of multiplication. The product of 3 and 5 is 15 because $3 \times 5 = 15$. An expression may be called a product if the terms are to be multiplied. The product of a and b is ab. The symbol $\prod$ is the product counterpart of $\sum$, the summation symbol. In notation, $\prod_{n=1}^{10} n = 3628800$ means the product of all the numbers from 1 to 10, $1 \times 2 \times 3 \times 4 \times 5 \times 6 \times 7 \times 8 \times 9 \times 10$, is 3628800.

Quadratic equation A quadratic equation is a polynomial equation with the maximum exponent of 2 and can usually be reduced to the form $ax^2 + bx + c = 0$, where a, b and c are any numerical coefficients, except that $a \neq 0$. The two solutions for x are $x = \frac{-b \pm \sqrt{b^2 - 4ac}}{2a}$. For example, if the equation is $2x^2 - 5x + 3 = 0$ then the solution is $x = \frac{-(-5) \pm \sqrt{(-5)^2 - 4 \times 2 \times 3}}{2 \times 2} = \frac{5 \pm \sqrt{25 - 24}}{4}$. So x is 1 or $\frac{3}{2}$.

Quotient A quotient is the outcome of the division of a dividend by a divisor. For example, in $\frac{10}{2} = 5$, 10 is the dividend, 2 is the divisor and 5 is the quotient.

Ratio A ratio expresses the relationship between two numbers of the same kind. For example, the number of times a bicycle pedal has to be turned for each turn of the back wheel is the ratio of the number of teeth on the circle attached to the back wheel to the number of teeth on the pedal circle. Ratio is written in a form of 3:1 which in this case would mean that there are three teeth on the back wheel circle for every tooth on the pedal circle.

Rational and irrational numbers Rational numbers are those that can be expressed as a whole number such as 27, or as a terminating decimal such as 1.234, or as a fraction itself made of rational numbers, such as $\frac{5}{11}$. Real numbers that cannot be so expressed are irrational, such as π, $\sqrt{2}$ and $\log_{10}3$.

Real numbers Real numbers are those numbers that extend positively from 0 in the direction of infinity and negatively from 0 towards minus infinity. Real numbers include whole numbers or integers such as 7, non-integer rational numbers such as $\frac{3}{7}$, and irrational numbers such as π, e and $\sqrt{3}$. See **integer** and **rational and irrational numbers**.

Significant figures When a number is quoted to so many significant figures that means it is to the prescribed number of digits counting from the first non-zero digit. The number 1.234567 quoted to three significant figures is 1.23 because, starting with the 1, there are three figures. 12345 to three significant figures (s.f.) would be 12300; 0.012345 to three s.f. would be 0.0123.

Simultaneous equations A set or system of simultaneous equations in two or more variables consists of equations that are all valid at the same time. If the equations are linear, that is, contain only the individual variables multiplied by numerical coefficients and numbers related only by addition and subtraction, they may be solved by subtraction or substitution methods. For example, if the equations are $x + 2y = 8$ and $3x + y = 9$ then, using the method of substitution, make x the subject of the first equation $x = 8 - 2y$ and then substitute for x in the second equation to give $3(8-2y)+y=9$. This simplifies to $24 - 6y + y = 9$ and so $-5y = -15$ and hence solve as $y=3$. Next put 3 as the value for y into either of the original equations (the first one here) and so solve for x. $x+2\times 3=8$, hence $x=2$.

Slide rule A slide rule is an analogue device used for computing. The principle is that two distances can be added by moving the centre part of the rule. If the distances are proportional to the logarithms of numbers then the effect is to multiply the two numbers concerned. Thus, a multiplication sum has been converted to a simpler addition sum. For example, in the picture showing a detail of a slide rule, it is set up to multiply by 1.25. The value of 1 on the C scale is put (by sliding the scale) over the position of 1.25 on the D scale. Decimal points and orders of magnitude must be imagined. Now any point on the D scale is 1.25 times its counterpart on the C scale; for example, $1.25 \times 12 = 15$. Slide rules are seldom seen nowadays as calculators have greater precision. A slide rule cannot give an answer to greater than three, or at best, four significant figures. However, a slide rule in practised hands is probably just as quick as a calculator. See also **logarithm** and **log table**.

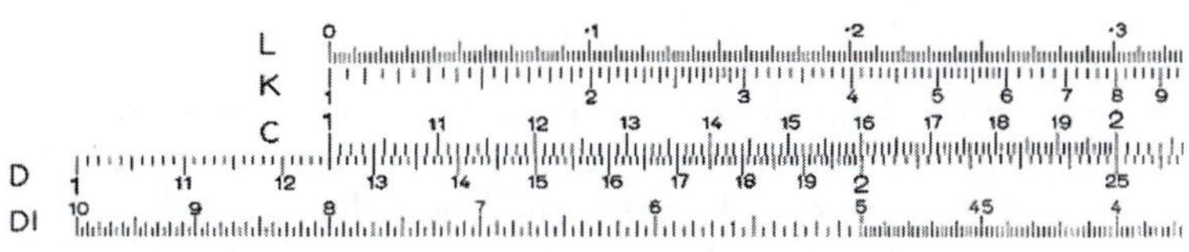

Slope, positive and negative The slope of a line is its gradient, which is also the tangent of the angle it makes with the x-axis. A positive value for the slope indicates that it is rising from left to right and a negative value indicates it is falling from left to right. A line with a slope of zero is parallel to the x-axis. A line parallel to the y-axis has an infinite slope. In the equation for a straight line $y=mx+c$, m is the slope of the line.

Square root The square root of a number is that number which when multiplied by itself gives the original number. Four is the square root of 16 because $4 \times 4 = 16$. It is also true that $(-4) \times (-4) = 16$, which implies that –4 is also a square root of 16. Whether it is or not depends upon the application but, by default, the square root normally means the positive square root, 4 in this example.

Standard deviation Standard deviation is a measure of dispersion or how a variable may depart from its mean value. For example, if a large number of people independently measure the distance between two points along a road about a mile apart as accurately as they can, it is unlikely that any two will come up with exactly the same answer. An average value could be found by taking the mean of all the values.

The standard deviation at its simplest is found as $\sigma_x = \sqrt{\frac{\sum(x_i-\bar{x})^2}{n-1}}$, where x_i are the measurements, $\bar{x}$ is the mean of the measurements and n is the number of measurements. This is only valid if there are no rogue measurements in the set. A small value of the standard deviation indicates good consistency among the measurements, that is, that the dispersion of the measurements is small.

Summation Summation is the addition of terms in a sequence. In notation, $\sum_{n=1}^{10} n = 55$ means the sum of all the numbers from 1 to 10 is 55.

Transposing a formula A formula may be transposed to make (one of) its independent variable(s) become the dependant variable. For example, the formula for the circumference of a circle, c, may be expressed as $c = 2\pi r$, where r is the radius of the circle. By dividing this equation by 2π we get $r = \frac{c}{2\pi}$ and the equation is said to have been transposed to make r the subject, that is, the dependent variable.

Trigonometric identity A trigonometric identity is an equality that relates trigonometric functions of one or more variables. A trigonometric identity is true for all possible values of the variables.

Trigonometric ratio Trigonometric ratios are ratios of the sides of a given right-angled triangle and are related to the angles of the right-angled triangle. For a given (non-right) angle the sides involved are those opposite the angle, adjacent to the angle and the hypotenuse which is opposite the right-angle. Taking any two of these three sides with respect to the given angle, in ratio form, leads to the full set of trigonometric ratios as:

$$\text{sine} = \frac{\text{opposite}}{\text{hypotenuse}}, \qquad \text{cosine} = \frac{\text{adjacent}}{\text{hypotenuse}},$$

$$\text{tangent} = \frac{\text{opposite}}{\text{adjacent}}, \qquad \text{secant} = \frac{\text{hypotenuse}}{\text{adjacent}},$$

$$\text{cosecant} = \frac{\text{hypotenuse}}{\text{opposite}}, \qquad \text{cotangent} = \frac{\text{adjacent}}{\text{opposite}}.$$

Truncation Truncation is a method for reducing the apparent precision of a number by cutting off its least significant digits, but ***without*** rounding. For example, truncating 1.23456 to 3 decimal places gives 1.234 whereas rounding to 3 decimal places gives 1.235.

Index